BANNERS OF THE TEUTONIC ORDER AND THE POLISH-LITHUANIAN ARMIES AND THEIR ALLIES AT TANNENBERG, 1410

From the *Banderia Prutenorum* and the *Annales* of Jan Długosz (1415-1480)

Illustrated using Flat Tin Figures

Originally published as:

Banner des Deutschen Ordens und der polnisch-litauischen Armee und ihre Verbündeten bei Tannenberg 1410

Richard L. Sanders
Colonel, U.S. Army Reserve, Retired
M.A., M.S.

Dedication

This book is dedicated to Werner Kästner (†) and Hans Jörg Stoll (Überlingen, Germany) in recognition of their achievements in creating their lovely standard-bearers for the 1410 Battle of Tannenberg as well as to Günter W. Fricke (Ilsenburg) for his outstanding research and publications on the topic which he generously allowed me to use.

© by Richard L. Sanders, Richmond, Virginia, 23225, USA
   and
Manfred Levec, Sindelfingen, Germany,
both as publishers.

All rights reserved. No part of this booklet may be reproduced, edited, stored or translated electronically, photomechanically or in any other form without the permission of the copyright holders.

Author: Richard "Rick" L. Sanders, Richmond, VA, USA

Editor: Manfred Levec, Sindelfingen, Germany

Graphics: Peter Bunde, Herzogenrath, Germany

Printing and production: On Military Matters, Hopewell, NJ

Originally published as "*Banner des Deutschen Ordens und der polnisch-litauischen Armee und ihre Verbündeten bei Tannenberg 1410*"

Edition: April 2024

CONTENTS

Acknowledgements	4
Foreword	5
The Teutonic Order at the 1410 Battle of Tannenberg/Grunwald	7
The Teutonic Knights	7
Flags of the Order's Army	7
The *Banderia Prutenorum* and Later Sources	9
The Flat Tin Figures	11
The *Banderia Prutenorum* as the Source for the Flags of the Teutonic Order and its Allies	12
The Banners in Detail	13
Possible Order's Banners Not Captured at Tannenberg	42
The Livonian Banners Captured in 1431	45
Banners of the Polish-Lithuanian Army at Tannenberg	48
The Polish-Lithuanian Army and its Disposition at Tannenberg	48
The Sources on the Polish-Lithuanian Army	50
Polish Heraldry	51
The Flat Tin Figures	53
The Banners of the Royal Polish Army	53
The Lithuanian Contingent and its Banners	72
The Ruthenian Contingent and its Banners	74
The Tartars and Their Standards	76
In Conclusion	77
Annexes	
1 Banners of the Teutonic Order's and its Allies for Attaching	78
2 Teutonic Order Banners Produced by Hans Müller (Erfurt) 1930-1938	86
3 Teutonic Order Banners from the *Banderia Prutenorum*, etc.	87
4 Günter Fricke's Map: The Polish-Lithuanian Kingdom in 1400	89
5 Polish Banners from the *Banderia Apud Grunwald I* Book	90
6 Polish-Lithuanian Flags by Erwin Ortmann	95
7 Günter W. Fricke's Polish-Lithuanian Banners and Those of Their Allies at Tannenberg	96
8 Polish-Lithuanian Coats of Arms after Günter W. Fricke	101
9 Polish Coats of Arms from the *Armorial of Knights of the Golden Fleece*	104
10 Polish Coats of Arms from the *Codex Bergshammar*	107
11 Contributors	111
12 Sources and Recommended Reading	112
End Notes	115

ACKNOWLEDGEMENTS

My special thanks to Bernt Grimm (Borlange, Sweden) whose willingness to paint so many figures made the book a possibility. My thanks also to Gianpaolo Bistulfi (Milano, Italy), Peter Blawatt (Saalfeld, Germany), Dirk Ewert (Bochum), Horst Meißner (Drößnitz), Gernot Schlager (Leonding, Austria), Jan Sennebo (Helsingborg, Sweden) and Stefan Wachter (Kulmbach) for the photos of their beautifully painted figures and to Wolfgang Friedrich (Rackwitz) for the photos of his figures painted by Jörg Hensel. Wolfgang Büche (Halle/Saale) and Dr. Thomas Brümmer (Halle/Saale) get special thanks for their information about the flags as does Peter Bunde (Herzogenrath) for his help with the graphics. My thanks also to Rolf Fuhrmann for the use of his flag plates and to Günter W. Fricke (Darlingerode) for the use of his drawings and article. I am especially grateful to Hans Jörg Stoll for his generous contribution of figures for the project. And my gratitude to my wife Ellen for her patient support of this effort.

My renewed thanks to Manfred Levec (Sindelfingen) for his editing, support with the research and the obtaining of figures.

The Great Banner of the Teutonic Order and Grand Master Ulrich von Jungingen
Figures designed and engraved in the 1930s by Ludwig Frank and produced by
Hans Müller (Erfurt) (left MP25, right MP24). - Painted & photo by Dirk Ewert

FOREWORD

This work deals with the banners of the Teutonic Order's army and those of the Polish-Lithuanian army and both sides' allies who fought at the 1410 Battle of Tannenberg/Grunwald that led to the decline of the Teutonic Order. This book uses flat tin figures painted by the author and Bernt Grimm, as well as other collectors, to illustrate the standards and flags. It attempts to use flat tin figures that were explicitly created to represent German, Polish, Lithuanian, Ruthenian (northwestern Slav) units and Tartars but sometimes had to resort to others that could be painted to illustrate these banners.

This book is mainly based on a selection of medieval to modern sources. Fundamental to this effort are the works of the Polish chronicler Jan Długosz (1415-1480) – his *Banderia Prutenorum* (Prussian Banners) and his *Annales seu Cronicae incliti regni Poloniae* (Annals or Chronicle of the Illustrious Polish Kingdom - written 1455-1480). In the middle of the 19th century, the Prussian historian F. A. Vossberg wrote his detailed work *Banderia Prutenorum oder die Fahnen des Deutschen Ordens und seiner Verbündeten, die in Schlachten und Gefechten des 15. Jahrhunderts eine Beute der Polen wurden* (Prussian Banners or the Flags of the Teutonic Order and its Allies in the Battles and Fighting in the 15th Century that became the Poles' Trophies). A critically important academic work on the banners was the Swedish historian Jan Ekdahl's 1976 *Die "Banderia Prutenorum" des Jan Długosz - eine Quelle zur Schlacht bei Tannenberg 1410 – Untersuchungen zu Aufbau, Entstehung u. Quellenwert der Handschrifts* (The "Banderia Prutenorum" of Jan Długosz – a Source on the 1410 Battle of Tannenberg – Investigation of the Manuscript's Composing, Development and the Value of its Sources).

Title page of F. A. Vossberg's *Banderia Prutenorum – die Fahnen des Deutschen Ordens und seiner Verbündeten*

At the end of the 20th century, Heinz Schenzle (*8 March 1923 -†9 June 1988) drew on the research of Günter W. Fricke (Darlingerode, GDR), and with the help of Andrzej Klein (Łódź, Poland), published an article and a booklet, both in German, about the Teutonic Order's and Polish-Lithuanian Armies. In 2000, the most recent important books on the banners at Tannenberg were published by Andrzej Klein. Those books contained not only the banners of the army of the Teutonic Order, but also those of the combined Polish-Lithuanian army and its Ruthenian allies. Klein's books are the *Banderia Apud Grunwald I: Chorągwie polskie pod Grunwaldem – Polish Banners at Grunwald*, co-authored by

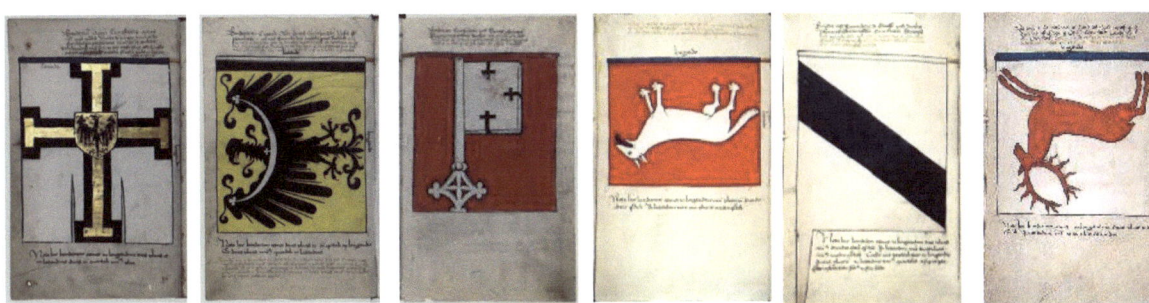

Banners from the 1410 Battle of Tannenberg, extract from the *Banderia Prutenorum*

Nicholas Sekunda and Konrad A. Czernielewski and the second volume, with coauthor Piotr Nowakowski, *Banderia Apud Grunwald II: Chorągwie krzyżackie pod Grunwaldem – Teutonic Banners at Grunwald* published by Alexander Co. in Łódź, Poland. Andrzej Klein's more recent volumes have benefited from specialist translations of Jan Długosz's original Latin works, as well as more than 15 years of additional research plus all the information available on the Internet. Unfortunately, all these books are out of print, which was another reason to create this publication.

Banderia Apud Grunwald I
by Klein / Sekunda / Czernielewski

Banderia Apud Grunwald II
by Klein & Nowakowski

It is my intention to provide an overview of the bulk of the flags that flew over the two gigantic armies that met at Tannenberg or Grunwald, as the Poles call the battle's location, and to illustrate it with many of the various flat tin figures which Bernt Grimm, other collectors and I painted.

Rick Sanders
Candler, North Carolina, USA
June, 2023

   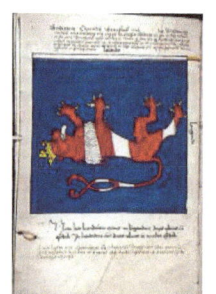  

Banners from the 1410 Battle of Tannenberg, extract from the *Banderia Prutenorum*

THE TEUTONIC ORDER AT THE 1410 BATTLE OF TANNENBERG/GRUNWALD

On 15 July 1410, the decisive battle took place in the area of Tannenberg/Grunwald in today's Poland, the outcome of which meant the end of the Teutonic Knights' dominance in northern central Europe. The combined armies of Poland and Lithuania, supported by Ruthenians (Slavs) and Tartars, decisively defeated the army of the Teutonic Order.

How the Teutonic Order knights, their "guest" knights and the Polish knights appeared in this battle was characterized by the bascinet helmet, with or without a visor, arm and leg protection of plates, lamellar armor, and short surcoats, mostly with no shields and no caparisons. The transition to plate armor only took place belatedly in Germany and the bordering lands.

Teutonic Knights

The Teutonic Order's army was made up of senior dignitaries (*Grossgebietigern*) like the Grand Master (*Hochmeister*) and the Order's Marshal among others, "Commanders" (*komthuren* in Middle High German), lesser Teutonic Knights, "*Sarjanten*" and further "servants" (*Dienern*) who belonged to the Order. The Teutonic Knights only made up a small portion of the Order's army at the Battle of Tannenberg. Although many Teutonic Order flat tin figures are beautifully drawn and engraved, many are now thought to be historically inaccurate. For example, the heavy horse armor is questionable and it is highly doubtful whether the Teutonic Knights wore their white cloaks in battle because they certainly would have restricted the combatants' movement. Interestingly, the Order's war and service horses were primarily black according to contemporary documents. The horses' black coloring was especially common in those from the area around the Vistula Lagoon and the Vistula Delta.[1]

"*Der Tannhäuser*" as a Teutonic Knight, from the *Codex Manesse* (probably produced between 1290 and 1340)

It should also be noted that the Order's "commandery" (*Komturei*) was an administrative district that was headed by a "*komthur*" or "*Komtur*" with a "*konvent*" (a convent ideally consisting of 12 Knight Brothers). The *Komtur*s' seats were usually in a castle belonging to the Order, an *Ordensburg*. In wartime, the *Komtur* led the commandery's levy under its own banner (flag). It is also clear that an element subordinate to a commander, i.e., a steward (*Vogt*) or stewardship (*Vogtei*) or in some cases even a caretaker (*Pfleger*) or his charge (*Pflegeamt*), could have its own flag.

Flags of the Order's Army

From the Order's ledgers, it is known that the vast majority of its flags were made of silk with emblems painted onto them. There were also small flags and pennants for tents or wagons that could be lined with stiffening material. For example, one of the Order's accounting documents recorded "*And 5 marks for Peter the painter for painting the Grand Master's banner, as he wants to leave on an expedition.*" (1399).[2]

On the medieval battlefield the flags, along with the war cries, were the most important means of recognition and communication. While the Order's Brothers were easily recognizable by their white attire, there are no indications that those other men who were obligated to serve with the Order, guest knights or mercenaries affixed even a small symbol associated with the Order to their clothing or armor. This was in contrast to the Polish-Lithuanian army which used a visible device. So for the Order's army, the flags constituted the primary visual means for command and control of the forces.

On the battlefield, banners (flags), along with serving to identify the units gathered beneath them, also had the practical function as a means of signaling. Certain kinds of waving, raising or lowering the flag served as a way to convey orders to the troops. Lowering one's own flag acknowledged defeat and also was the prompt for the combatants to flee. For example, Nitze von Renis, the standard-bearer of the *Kulmer Land*'s[3] contingent at the 1410 Battle of Tannenberg/Grunwald was beheaded for treason because he – apparently with treasonous intent – had lowered the banner and thus called for their troops to flee the field.[4,5]

There is strong evidence that a commandery had multiple flags. For example, according to the contemporary *Ämterbuch*[6] (official record) inventories, in 1392 the Brandenburg Commandery had: "*30 ells* of tablecloth, also 2 red and 1 yellow for banners (banyrs).*" In 1412 it had "*4 banners, also 11 ells of yellow* [cloth] *for banners.*" In 1400, the Ragnit Commandery had "*6 red banner cloths, also 25 ells white linen, also 3 large new banners.*" In 1407 it had "*30 ells of banner linen.*" In 1414: "*3 complete banner cloths, 3 main banners* (main banner: *hewptbanyr, howtbanyr*), 4 *Witting's banners* (*wittingisbanyr***), 35 marking flags (sticbanyr)." In 1419: "*2 main banners, 2 Wittinge's banners, 24 vanguard flags (rennerbanyr).*"***

Similar to the Order army's very diverse composition, the army's various elements carried a colorful variety of flags. "Gonfanons" and "banners" can be distinguished within the general designation of "flags." The gonfanon is usually a narrow and elongated flag that has its device arranged perpendicular to the staff, so that to the observer it looks sideways. Gonfanons ended in several streamers or tails. Banners are rectangular or square, and the device is displayed parallel to the staff, appearing upright to the observer.

Because the gonfanon is the older type of flag, it is ranked higher than a banner. The Marshal's flag, which was equivalent to the Order's overarching flag, and the Grand Master's flag were gonfanons. Their having had slits are indications of the original form of the gonfanon having tails, and the axis for the emblem is horizontal. Except for the appearance of the flags from those passed down in depictions, very little is known about their meaning, their respective ranks, their function and uses within the Order.[7]

One of the Order Army's main standards was the "*renne-vane*" or "*rennerbanyr.*" In Middle High German (MHG) "*rennen*" meant to run or ride quickly. This standard was carried in the troop that held the most honored position in the front line of a medieval army, i.e., the vanguard. This troop itself also often was called the "*renne-vane.*" Certain flags had precedence over others. The Virgin Mary flag and St. George flag were senior to all other flags. Then followed the Grand Master's banner and the Marshal of the Order's banner that also served as the Order's overarching banner. These banners had a superordinate function so that other banners could rally to them. The banners of the Ragnit and Insterburg Commanderies and the *witing's* banner had the privilege of "*vorstrît,*" (the right to precede), as in the honor to be the first to attack the enemy.[8]

---

*An "ell" is described below in the section titled "The *Banderia Prutenorum* and Later Sources" on page 9.
**A *witing* (pural: *witinge*) was the Middle German term for an ethnic Pruss official in the Order's service.
*** NOTE: The "*sticbanyr*" may have been small flags or pennants intended to mark paths for marching troops or to be used within the field camp. The "*rennerbanyr*", because of their great number mentioned, in this case were probably made for use by advance guard detachments.

Depending on the high-ranking guest knights present, the order of rank could vary. For example, when Duke Albrecht III of Austria was a guest on the Order's 1377 expedition, the succession was first the Ragnit Commandery, followed by St. George, then the Austrian banner of Steierland (Styria), then the Grand Master's banner and finally the Duke's banner of Austria.[9] The Saint George Banner will be discussed in detail on page 20 and the Banner of the Virgin Mary in the section on the banners captured by the Poles after the Battle of Tannenberg pages 21 and 45.

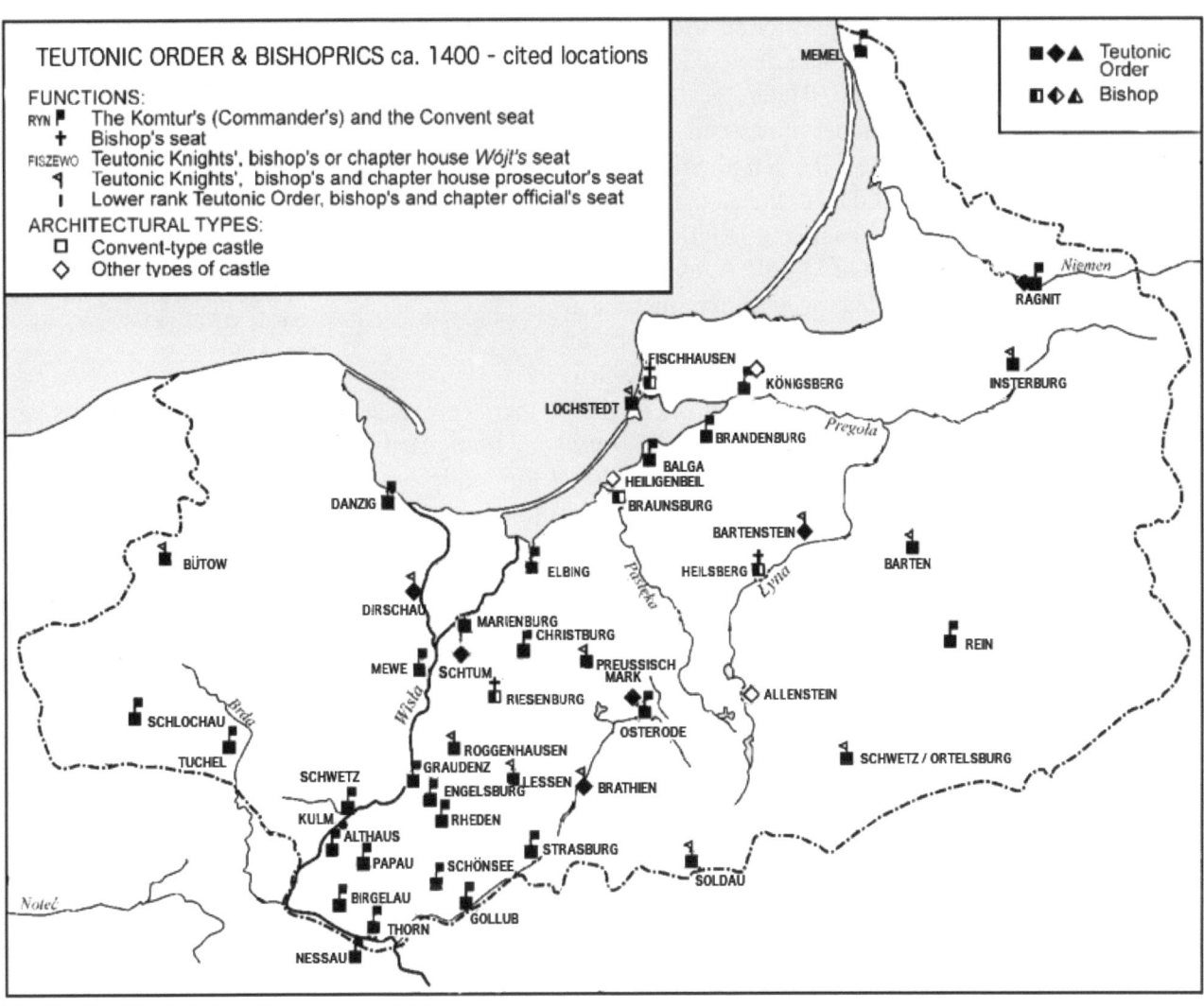

The *Banderia Prutenorum* and Later Sources

The key source on the standards of the Teutonic Order and its allies at Tannenberg is the *Banderia Prutenorum*. The manuscript, on 48 parchment sheets, 18.6 × 29.3 cm, (7 3/8 x 11 9/16 inches) was written by Jan Długosz (*1415; † 1480) and illustrated by Stanisław Durink. It depicts 56 Teutonic Order's army vexillae or banners of which 51 are from the Battle of Tannenberg, one from the Battle of Koronowo (German: *Polnisch Krone*, from the Schwetz Commandery or the Steward of Neumark) later in 1410 and the last four banners from the 1431 Battle of Dąbki (German: *Nakel*). It should be noted that not all of the Order army's banners were captured by the Poles as part of the Order's army fled to the Marienburg and any flags they carried therefore could not be depicted in the *Banderia*. It is also important to mention that only 46 of the banners were initially found in the *Banderia*. Ten were only added later, not so aesthetically done and on the reverse (right side) pages of the existing sheets, so they are referred to as the *Rectobanners*.[10]

The *Banderia*, written in Latin, identifies who supposedly carried the banner, the organization it represented and the circumstances of its capture as far as known or believed by the Poles. But the accuracy of much of this information is questioned today. Each page starts with a title of the banner depicted. For most of the flags, other than those added later, the description includes the size of the banner measured in Polish "ells" (cubits). The "ell" was the length of a man's forearm, sometimes measured from the elbow to the tip of the middle finger. In any case, the length of an ell varied from town to town in the Middle Ages, so that it is not possible to get an exact size in centimeters or inches. For our purposes, it will suffice to calculate an ell on average as being between 47 and 52 cm. So it is at least possible to determine the relative sizes of the banners compared to one another. The flags varied in size from the smallest, the Grand Master's "*Rennfahne*" (vanguard flag) (1¼ ell x 1 ell) to the largest from the city (*Stadt*) Kulm (3 ells x 3 ells and a 3¼ ell long tail). So the smallest flag was 59 to 65 cm long and 47 to 52 cm tall; the largest was 141 to 156 cm long and wide with a tail that was 153 to 169 cm long.

Extract from the *Banderia Prutenorum* showing the banners of the Bishopric of Pomesania and of the Commandery and Town of Graudenz

Many descriptions of the banners in the *Banderia Prutenorum* are based on publications from the 19th century, for example that by F. A. Vossberg from 1849,[11] and those works identify inaccuracies. As mentioned above, the *Banderia Prutenorum* is a manuscript from the 15th century with illustrations of flags that were captured by the Poles at the Battle of Tannenberg and two later conflicts they fought against the Order. So it is understandable that the Poles did not correctly identify a number of these captured flags. Based on current knowledge, it is clear that many of Długosz's designations are wrong; however some of these errors have tenaciously remained to this day.

From left to right the Grand Master's *Rennbanner*, banners of the Graudenz & Ragnit Commanderies, of the Brattian Stewardship & of the town of Heiligenbeil – Figures by Werner Kästner, painted & photo by R. Sanders

The Flat Tin Figures

There are many 30mm flat tin figures (German *Zinnfiguren*) that were specifically created or are usable for the 1410 Battle of Tannenberg, and many of them have integral banners or flags. Especially well-known among them are the figures by Hans Müller of Erfurt, but there are also castings from Wolfgang Friedrich, and the former Golberg International firm (now available from Wilfried Dangelmaier) among others. Werner Kästner produced 40 Teutonic Order flat tin figures with engraved banners for Tannenberg (now available from Schmalkaldener Zinnfiguren[12]). Hans Jörg Stoll produced both Teutonic Order and Polish-Lithuanian standard bearers, with and without engraved devices on the flags. Additionally there are various editors who offer Ruthenian and Tartar standard bearers, some of which will be described later.

Hans Müller's MP29 Banner of the Commandery and Town of Engelsburg
Painted & photo: Gianpaolo Bistulfi

There is no question that most of the flat tin figures with the flags and banners of the Order, its commanderies and allies are based on examples from the *"Banderia Prutenorum"*, probably drawn from the 1849 work of F. A. Vossberg (see page 5).

Yet in some cases the figures' banners are too large (like those by Golberg and some by Kästner) or the flag staffs are not strong enough so that they tend to bend. Hans Müller (Erfurt) only produced two Teutonic Order figures with banners – the "Great" Grand Master's banner and the banner of the Engelsburg Commandery, but he created an additional 24 tin flags of the Order's army as ones that can be attached to lances (see Annex 2). Fewer and fewer collectors are prepared to use such attachable metal flags but there are alternatives. Among them are flags made of paper that can be glued, like the ROFUR flags by Rolf Fuhrmann or the ones presented here in Annex 1.

The ROFUR Flags offer an extensive series of paper flags of the Teutonic Order, the Polish-Lithuanian units and their allies, in 1:72 scale, that can be cut out and attached to lances or flag staffs (see illustration below). These flags are available through the Internet at http://www.rofur-flags.de.

While this book focuses only on the flat tin figures (Zinnfiguren) suitable for depicting standard-bearers for the Teutonic Order, its opponents and their allies, there are of course a number of 3-D 28mm wargaming figures by companies like Perry Miniatures and Foundry with products suitable for this as well, and some of the flags in the appendices are close enough in scale to be used with them.

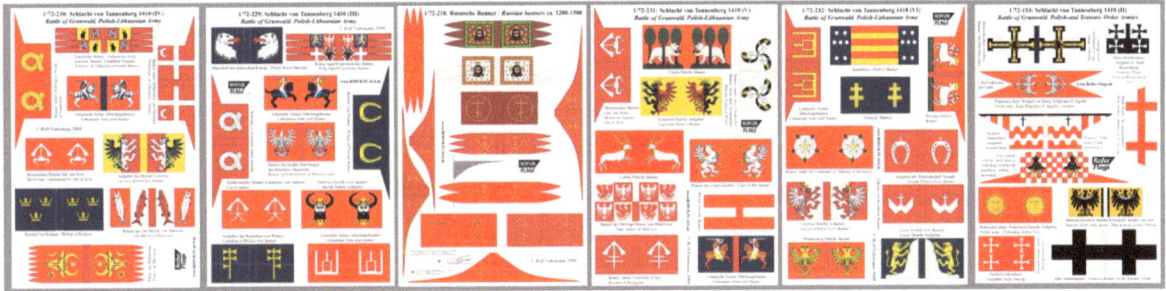

Various Tannenberg/Grunwald 1/72 or 28mm Banners from ROFUR Flags

The *Banderia Prutenorum* as the Source for the Flags of the Teutonic Order and its Allies

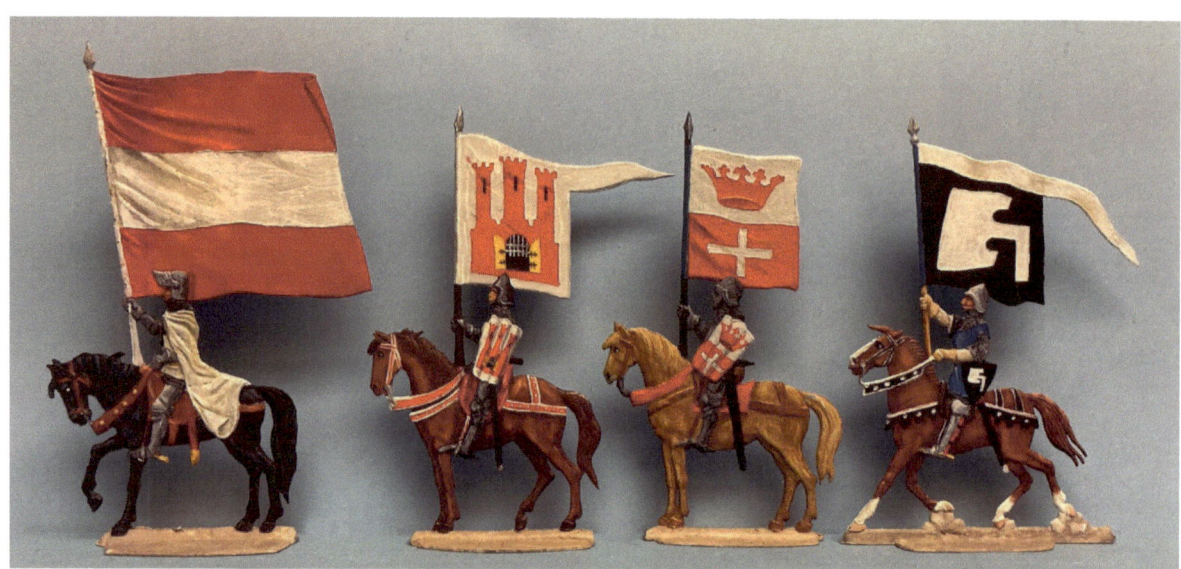

From left to right: Banner of the Grand Commander (*Großkomtur*), of the Town of Thorn, of Altstadt Königsberg & of the Bartenstein Chamberlain's Office - Figures by Werner Kästner, painted & photo by R. Sanders

The banners of the Order's army can be classified as representing the Order's dignitaries (*grossgebitiger*), allied bishoprics, the Order's commanderies (*Komtureien*) and Order stewardships (*Vogteien*), often together with towns or cities, towns/cities not under commanderies, allies and various consolidated units. The leaders or dignitaries included the Grand Master, the Marshal of the Order, the Grand Commander, the Senior Treasurer and the Senior Draper among others, most of whom were killed at the battle. The

Duke of Stolpe or Stettin
Golberg Go2903 (Wilfried Dangelmaier)
Painted & photo: B. Grimm

Prussian bishoprics that sent their troops to the Order's army and whose banners were captured were Pomesania, Kulm, Sambia (*Samland*) and Warmia (*Ermland*). The commanderies, as the Order's administrative districts, sent their own banners (shown below). Among the levies from cities and towns were, according to Długosz, Allenstein, Bartenstein, Braunsberg, Heiligenbeil, Kulm and Thorn whose flags were captured by the Poles. The allies included the Dukes of Oels and Stettin as well as a levy probably from Hungary (red banner with white cross), various "guest knights" from outside the Order's state of Prussia who fought under the St. George Banner, along with mercenary units.

The designations of the flags in Annex 1 reflect, as much as possible, the current knowledge. For example, both the banners with the crossed arrow and crossbow bolt are assessed to have flown over consolidated units of archers and crossbowmen from various levies. The erroneous designations of "Banner of the Swabian Knights" and "Banner of the Commandery and Town of Mewe" go back to Długosz. It can be deduced that the flag of the Mewe Commandery would have been a version of the current coat of arms of the town Gniew in Poland – a seagull – also the seals of the Commander and the House Commander of Mewe show such a "canting" device (depicting the spoken name). Similarly Długosz's designation of the flag of the

Schlochau Commandery is incorrect since the seal of the Schlochau Commandery displays a recumbent bull with a cross of the Order.

Długosz identified banners as belonging to units from Westphalia, Swabia or Switzerland; however, the Teutonic Order's registries that have been published in recent years prove that no knights from those three regions took part in the battle. We do find mercenary detachments that fought in the battle: the bands (*Rotte*) of Nickel von Kottwitz, of Kaspar von Gersdorff, of Heintze von Borsnitz, the Meissner *Rotte* and others. The mercenaries were dispersed among various banners (units) of the Order where needed. Only the "Meissners" (men from Meissen) fought under their own red and blue quartered banner.

As mentioned above, the *Banderia Prutenorum* only offers information about the banners and flags that were captured at the battle, so it is possible to extrapolate the appearance of a few other banners for discussion's sake. These attempts are based on contemporary impressions on seals of the individual commanderies, bishoprics or towns. One example is the banner of the Senior Hospitaller (*Oberster Spittler*) as depicted in a mural in the Order's castle in Lochstedt (now Pavlovo, RUS, Kaliningrad Oblast), which was not among the banners shown in the *Banderia*.

The banners in Annex 1 are taken more or less directly from the *Banderia Prutenorum* manuscript, but "mirrored" and re-sized to correspond to 30mm flats. They are intended to be cut out, folded over and glued onto a flagstaff or lance. Of course, one can also fold or bend them to look more realistic since no flag or banner would be straight and unwrinkled. It is possible to download and print each sheet with the flags at the following website: (https://www.ricksanderszf.com/1410-tannenberggrunwald.html).

The Banners in Detail

The following descriptions come primarily from the works of F. A. Vossberg and Andrzej Klein and are presented in the sequence that the appeared in the *Banderia Prutenorum*. The quotations are based on F. A. Vossberg's translations of the Latin text in the *Banderia Prutenorum*. The quotations from F.A. Vossberg's work are referenced with page number, plate number and image number in the Endnotes.

The Grand Master's "Great" Banner
Hans Müller's MP25 Halt, engraved.
Painted & photo: Gianpaolo Bistulfi

The **Teutonic Order's Grand Master's "Great Banner"** was described by Długosz thusly: "*The banner of the Master of the Crusaders, whom the Grand Master Ulrich von Jungingen led, among which were the most excellent courtiers and warriors... This banner had a length of three ells and a width of 2¼ ells.*" Vossberg wrote that the flag had a "white cloth with double slits, with a large black Latin cross, on the Latin cross of Jerusalem, decorated in the center with a golden shield on which rested a black eagle. The cloth on the staff is blue otherwise light yellow." It is noteworthy that in the depiction of this flag in the *Banderia* the shield is shown sideways, i.e., it is a gonfanon, while in Vossberg's book it is shown upright. Vossberg wrote further "*Based on coins and seals and other monuments from the 13th to 15th centuries, this consisted of two simple beams without the Latin cross addition, completely like the flag of the Senior Marshal*" (see p. 15). The figure by Hans Müller (Erfurt), engraved ca. 1938, shows the shield with the eagle vertical (above), while Wolfgang Friedrich's corres-

The Grand Master's "Great" Banner
Left: W. Friedrich's figure painted & photo by R. Sanders
Right: Werner Kästner's design (TB 03)

The Grand Master's Small Banner
Werner Kästner, engraved (TB 02)
Painted & photo by R. Sanders

ponding figure above and that by Werner Kästner (TB 03) show it horizontal like in the *Banderia* illustration.

Again, there is documented evidence that there were numerous copies of the Grand Master's banners and the Order's expense ledgers give the most extensive information about them. Standards with the Grand Master's emblem were used in numerous varying sizes and implementations. In 1400, the Treasurer's Book (*Tresslerbuch*) listed "*11 banners of the Master, each for 10 scot [monetary unit, 1/24 of a Mark], and 2½ marks for a flag for the Master...also paid 3½ marks for 5 banners to be fixed on the Master's tent and 16 marks for 2 flags...*" In 1409, payments to a painter were made for "*2 large silk flags with our Grand Master's coat of arms made with gold and 4 silk mid-sized flags also with our Grand Master's coat of arms with gold...4 small flags of cloth...13 very small flags painted with colors...*"[13] Additional small flags or guidons were used to identify the Grand Master's servants, to decorate the Grand Master's tent and his traveling wagon.

The **Grand Master's *Rennfahne*** (vanguard flag) or "Small Banner" was probably carried by the army's vanguard. According to the *Banderia*: "*Under this flag were the more excellent* (more knightly) *fighters of the Order, also some mercenaries from various regions of Germany, as well as some of the Master's Court and House members.*"[14] The flag cloth was white with a black Latin cross of Jerusalem, decorated in the center with a golden shield with a black eagle. The flag staff's sleeve was white, the staff dark yellow. "*This small banner ... was 1¼ ells long and 1 ell wide.*" The figure by Werner Kästner is still available.

Grand Master's "Great" Banner and his *Rennfahne* from the *Banderia Prutenorum*

Lochstedt Ordensburg fresco

Großkomtur banner from the *Banderia Prutenorum*

Grand Commander (*Großkomtur*)
Werner Kästner's (TB 04) engraved
Painted & photo by R. Sanders

In the *Banderia* it states *"The **banner of the Grand Commander** of Stuhm, led by the famous Teutonic Knight and Grand Commander Cuno von Lichtenstein, under him almost all Austrian mercenaries and a few Teutonic Knights. The banner is three ells wide and 3 ¾ ells long."* The flag's cloth consists of three horizontal stripes: one red, white and red. However, there is much controversy surrounding this flag. According to Vossberg, the flag corresponds to *"that of the Stuhm Stewardship (Allem) also possibly in the Stewardship of Nauen-Tober of the Prussian Order's officials: that of the then Steward of Stuhm, Heinrich Potendorf, who fell in the Battle of Tannenberg. In any case it appears that the flag shown more likely belonged to the Steward of Stuhm who also died in the battle than the honored Grand Commander Cuno von Lichtenstein. That Austrians were grouped around this banner perhaps was the reason that it is completely the same as the device of Austria itself."*[15] Vossberg's book on Prussian coins and seals showed a coat of arms for the Stuhm Steward with such a horizontal bar. Other scholars believed it was more likely the banner of Cuno von Lichtenstein.[16] In a similar way, Rolf Fuhrmann states in his book *"Der Deutschorden"* that this flag is the *"suspected banner of the Grand Commander and Commander of the Marienburg levy."*[17] It is also noteworthy that a red-white-red flag is depicted in a fresco in the Lochstedt Castle. However, that flag is triangular and with the device turned 90 degrees so the colors are parallel to the staff. It is shown along with the banners of the Grand Master and Marshal of the Order, implying that this banner belonged to one of the Order's senior officials. Stuhm is now Sztum, Poland.

The **Marshal of the Order's banner** was described by Długosz as follows: *"The Crusaders' banner led by Friedrich von Wallenrod, the Grand Marshal (Großmarschall) of Prussia, a Frank of an excellent lineage, whose shield bore a river with a cross and had a rooster as a helmet crest, was killed in this same battle and brought to the Marienburg Castle ... This banner had a length of 3¼ ells and width of 2¾ ells."*[18] This white flag was decorated with a black Order cross and was split into three parts at its end; the sleeve was white and black.

According to the Order's *Customs* (*Gewohnheiten*), Chapter 19: *"The*

Marshal of the Order (*Ordensmarschall*)
Figure & photo by W. Friedrich, painted by J. Hensel   Werner Kästner's design (TB 05)

15

bearer of the Marshal's banner shall be a Turcopole."[19] This was the case on the march and other occasions, while in the attack, the Senior Marshal carried the Order banner himself (*Customs*, Ch. 61).

Ordensmarschall's seal (Vossberg, *Münzen und Siegel*, plate I)

As Chapter 19 of the *Customs* originated in the era of the Order's activities in the Holy Land, the "Turcopole" in later times was obviously a squire rather than a Middle Easterner. Around 1407 a new seal showed the Marshal's flag with a long piece of cloth projecting from the top edge of the fly, called a "*schwenkel*" in MHG. The word "*Schwenkel*" derives from "to wave," and is the outermost end, tail or point of the main flag.

The Marshal of the Order was also the House Commander (*Hauskomtur*) of Königsberg and had supervision of the Order's military system. His seat was in Königsberg (present Kaliningrad, Russia).

Werner Kästner also produced a figure with the Marshal of the Order's flag engraved (TB 05).

The Order's Treasurer (*Ordenstressler*)
Left: H. Müller's "e" on his MR137), painted by M. Kröbel; photo by J. Schwarz; collection of H. Schwahn - right: Werner Kästner's design (TB 07)

The flag of the **Ordenstressler**, the treasurer and head of the central financial administration at the Marienburg, was described by Długosz as: "*The Crusader's banner led by Thomas Merkheim, the Ordens-Tressler, who was killed in the same battle, had many mercenaries and House members under the sign of his office.*"[20] This red flag bore a white key, had three cross-shaped cuts that were shown in black. The *Banderia* gives no size for the flag, as it is a "*Rectobanner*" added to the work after 1448. A figure with an engraved *Ordenstressler* flag (TB 07) was produced by Werner Kästner shows the device correctly as a gonfanon but adds slits, while the attachable flag by Hans Müller has the key device turned sideways in contrast to the depiction in the *Banderia*.

Ordenstressler's banner in the *Banderia Prutenorum*

Possible pennon of *the Oberster Spittler* in the Lochstedt Castle

Oberspittler's seal, 1347-1449 (Vossberg, *Münzen und Siegel*, plate I)

One of the banners of the Order's dignitaries depicted in a mural in the Order's Lochstedt Castle, but not shown in the *Banderia,* is assessed to be that of **the Senior Hospitaler** *Oberster Spittler)*. At the time of the battle, Werner von Tettingen, who was the Senior Hospitaler as well as the *Komtur* of Elbing, rescued himself from the defeat and remained in office until August 1412. It could well be that his banner was not among the banners shown in the *Banderia Prutenorum* because he was able to take it with him when he departed the battlefield. In the Lochstedt Castle's rather age-worn mural, it appears as a triangular, pennant-like flag, the upper half black, the lower half white and the sleeve similarly colored.

Pennon of the *Oberster Spittler* - H. Müller's MR132 with pennon added. Painted & photo: R. Sanders

So-called "Banner of the Bishopric and of the Bishop of Culm" (more likely of the Bishop of Sambia)
Left: H. Müller's MR129 with a flag from Annex 1, painted & photo by R. Sanders.
Right: Werner Kästner's *"Bischof von Samland"* design (TB 12)

The *Banderia* states *"The **Banner of the Bishopric and of the Bishop of Culm** (Kulm/Chełmno) was led by the bishopric's court man Dietrich von Sowemburg. Under him were the House and court members and vassals of the Bishop of Culm reigning at that time. This banner is 2¼ ells long and 1 7/8 ells wide."*[21] The flag had a white cloth with a red sword and bishop's staff crossed over one another; the sleeve was blue. However, according to Vossberg and more modern authors, it appears that Jan Długosz was confused because this coat of arms belonged to the **Bishop of Sambia (Samland)**. The Bishop of Sambia had his seat, starting in 1268, in a castle at Fischhausen (now Primorsk in Russia's Kaliningrad Oblast). The coat of arms of that town on a seal from 1305 clearly shows the crossed bishop's staff and sword as its device, as noted by Vossberg in his work on coins and seals.

Werner Kästner also edited a *"Bischof von Samland"* figure with an engraved flag (TB 12) but the device appears to have been reversed.

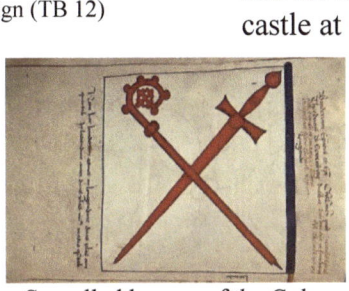

So-called banner of the Culm Bishopric in the *Banderia Prutenorum*

Fischhausen 1305 seal from Vossberg, plate XV

So-called "Banner of the Bishopric Ermeland"
H. Müller's MR129 with flag "h"
Painted & photo: R. Sanders

Another levy from a bishopric may have been sent to the battle by the Bishop of *Ermland* (Warmia). According to *Banderia*: "*Banner of the **Bishopric of Ermeland** (sic) alias Town of Heilsberg, led by [… consisted of?]…. House and Court members of the Bishop of Ermeland. Under him were great numbers of men born in the Bishopric of Ermeland, burghers from the town of Heilsberg and court and House members of the noted Bishop of Ermeland, which made up 100 lance-bearers if not many more. This banner is 2¾ ells long and 1¾ ells wide. But the tail is 1 ell wide where it starts at the top and becomes ever narrower.*" According to Vossberg, "*The red cloth, along with the tail, on the upper portion contains a lamb of God with a flag with a cross, standing in front of a chalice into which the blood pours from its heart. The lower, narrow field is white but the cloth on the shaft is red.*"[22] According to later research it is rather certain that this flag actually belonged to the Rehden Commandery. However, in the *Banderia* there is an illustration of a very similar flag, shown from the other side, that is identified as from the "*Commandery and Town of Schlochau*" (*Commendarie et Ciuitatis Slochow*) and is described as being smaller in ells – perhaps it was a levy's "reserve" (back-up) flag (see also page 39). However, this flag should not be discounted as representing the Bishop and Bishopric of Warmia because it closely resembles the seal of the town of Heilsberg (now Lidzbark Warmiński, Poland) where that cleric had his seat at times during the 14th and 15th centuries as evidenced in the Heilsberg seal from 1320.

So-called "Banner of the Bishopric Ermeland alias Heilsberg" in the *Banderia Prutenorum*

Heilsberg 1320 seal from Vossberg plate XVI

Bishop of Pomesania
Left: H. Müller's MR129 with his "n", painted & photo by Gianpaolo Bistulfi.
Right: Werner Kästner's *"Bischof von Pomesanien"* design (TB 11)

In the *Banderia* it states: *"The **Banner of the Bishop of Pomesania**, which Marquard von Kessenberg led, under him were gentry of the Pomesanian Bishopric and other mercenaries gathered by the Bishop of Pomesania."* According to Vossberg, *"A red cloth on which, between two golden bishop's staffs was the golden eagle of St. John, the head in a halo. On the white ribbon held by the eagle was written 'Scts iohannes' in black letters."* According to Długosz: *"This banner has a length of 2½ ells, but 2¼ ells in width."*[23] It should be noted that in the *Banderia* this flag is shown as a gonfanon with the eagle horizontal, which no figure has engraved that way (see below). Werner Kästner also produced a "Bischof von Pomesanien" figure with an engraved flag (TB 11). The bishop's seat was at the Riesenburg (Prabuty, PL).

The *Banderia* states *"The **Banner of the Bishop and Bishopric of Samland**, was led by Heinrich Grafen von* [Count of] *Kamenz von Meißen. Under him were vassals of the Samland Bishopric and other men of the convent and paid mercenaries whom the bishop led himself."* The banner is displayed in the *Banderia* from the left side and has a white cloth with three red hats arrayed under one another. *"This Banner is 2½ ells long and 2¼ ells wide."* According to Vossberg, this flag was exactly the same as that of the Ragnit Commandery (see page 37 below) and *"it is also certain that the flag which Długosz incorrectly attributed to the Bishopric of Culm (Kulm, Chelm or Chełmno) belongs to the Bishop of Samland."*[24] Jan Ekdahl suggests that this flag may have actually been the standard of the Order's steward (*Ordensvogt*) of Sambia, whose seal in 1322 showed a pointed cap.[25] The figure by Werner Kästner has the caps engraved on both sides.

So-called "Bishopric and Bishop of Samland"
W. Kästner's (TB 39), engraved.
Painted & photo by R. Sanders

Bishop of Pomesanien   So-called "Bishopric of Samland"
Extract from the *Banderia Prutenorum*

Senior Draper (*Obersten Trapier*)
Left: Golberg Go2899, painted & photo by R. Sanders.
Right: Werner Kästner's design (TB 06)

The Banner of the levy of the Christburg Commandery under the command of the **Senior Draper (*Oberster Trapier*)** was identified by Długosz as the "*Flag of the Grand Master's, Heinrich von Plauen's, and his Crusader Order.*" The Senior Draper was the administrator of the clothing system and the *Komtur* of the Christburg (Dzierzgon, PL). The flag's cloth was divided diagonally with the upper portion red and the lower portion white. According to Vossberg, the sleeve was blue but the flag sleeve displayed in the *Banderia* is white. According to Długosz, "*This banner is 2 5/8 ells long and 2¼ ells wide.*"[26] A second similar flag appears late in the *Banderia* and was identified as the "Banner of the Commandery and Town of Scythno [Schwetz]" or from "Ortelsburg" and it is presented below on page 36. It should also be noted that in Długosz's *Annales*, he writes that at a September 28, 1410 engagement near Koronowo (German name: *Polnisch Krone*) such a red and white diagonally halved flag was the standard of Michael Kuchmeister, who was coming from the Order's castle at Tuchel (Polish: Tuchola).[27] Werner Kästner produced, along with the standard-bearer shown here (TB 06), a "*Komturei Schwetz oder Ortelsburg*" figure with a similarly engraved flag (TB 38).

St. George Banner (*Georgsbanner*)
Left: H. Müller's "o" on MR165, painter: M. Kröbel; photo: J. Schwarz; collection of H. Schwahn. Right: Werner Kästner's design (TB 10)

It states in the *Banderia*: "*The **banner of Saint George** on the Crusaders' side, was led by the brave Georg Kerzdorf who considered it shameful to flee from the engagement. He remained unflinchingly until he was captured by Polish fighters and the flag was torn from his hands. Under him were famous knights from all parts of Germany, who were almost all killed courageously in the fighting, few were saved by fleeing. With their bodies, they died protecting the location where they stood.*"[28] However, the colors given in the *Banderia Prutenorum* appear to have been reversed and should have shown a white cloth with a red cross, if truly a St. George standard. Or this banner possibly belonged to a commandery, as Vossberg suspected.[29] The St. George banner (*Georgsbanner*) always displays a red cross in a white field. Długosz himself later reported this in his "*Annales*" (1455-1480) in which he wrote that the *"St. George banner bears the sign of a red cross in a white field."* The size of the flag's cloth is not mentioned in the *Banderia Prutenorum*, as this is one of the "*Rectobanner*" pages that was added later in the document's development. It has also been postulated that the colors are accurate but that it is a Saint Florian's banner instead of a St. George banner.

Further, according to Rolf Furhmann's book on the Teutonic Order, at the end of the 13th century, the Saint George Banner, in addition to the Banner of the Virgin Mary, took on more importance as a patron saint for the Order's Prussian branch. The banner with the Saint George's cross actually does not belong to the Order's flags in the closest sense. Next to the Virgin Mary banner (the latter primarily for the Order's Livonian branch), it was the most senior field standard under which the crusader army went to the field with the Order's army. However, occasionally the Saint George banner was also used when no guest knights were along.[30]

Until the mid-14th century, the version with the white cross on a red field, which was also the general war banner of the Holy German Empire, predominated during the crusaders' Prussian expeditions. The Germans, as well as the Dutch who belonged to the Empire, claimed carrying the imperial banner with the red field as their special prerogative wherever there were pagans to fight.[31] According to Paravicini, the imperial banner was last mentioned in 1378 with a silver cross on a red field.[32]. Ekdahl asserts it cannot be correct where Długosz describes the Order's guests from various German regions having fought under the red flag with a white cross, as the "Banner of Saint George," captured at the Battle of Grunwald. The Order's Saint George banner already had a white field with a red cross around that time.[33]

A figure with an engraved St. George banner was produced by Werner Kästner (TB 10, shown above).

So-called "Banner of the Swabian Knights"
H. Müller's "t" on his MR129
Painted by M. Kröbel, photo: J. Schwarz;
H. Schwahn's collection

The white flag with the crossed bolt and arrow in red, which Długosz identified as "**The banner of the Swabian Knights**" ("*Banner der schwäbischen Ritter*")[34] was certainly **an archer's standard**, as in a flag under which archers and crossbowmen of varying origin were gathered together in tactical fighting units. The erroneous designation as "banner of the Swabian Knights" (from *Banderium Militum Almanorum*) goes back to Jan Długosz. According to historian Wolfgang Büche, the publication in recent years of the Teutonic Order's accounting books proves that no knights from Swabia participated in the battle. According to the *Banderia Prutenorum*, "*This banner was two ells long and the same in its width.*" The author found no figures with a corresponding engraved banner.

Banners from the 1410 Battle of Tannenberg, extract from the *Banderia Prutenorum*

The so-called **"Banner of the Hessian Knights"** (*"Banner der hessischen Ritter"*) was described by Długosz thusly: *"The banner of the City of Braunschweig, which the Duke of Braunschweig led, as he led an army against the barbarians, had then been founded in Prussia and he had given it his name and that of his duchy. His and his duchy's coat of arms, namely a red lion, which was decorated in white stripes partially on its breast, partially on its stomach and partially on its tail, and decorated with a gold crown, had designated in a sky-blue field, was granted, dedicated and privileged. Its leader was the Commander, according to others the Steward of Braunschweig. Under him were Teutonic Knights, local men and burghers of the district and the city of Braunschweig. This banner is 2¼ ells long and 2 1/8 ells wide."*[35] The lion's crown and claws were in yellow, the tongue in white. Długosz certainly misidentified this so-called *"Banner der hessischen Ritter"* as being from Brunswick. Because the banner strongly resembles the Thuringian (and later Hessian) coat of arms, one can possibly conclude that Thuringian knights, rather than Hessians, fought under it, as the Thuringian nobility was heavily represented in the Order. Kästner produced a *"Banner der Ritter aus Hessen"* figure with an engraved flag (TB32).

So-called Banner of the Hessian Knights (*"der hessischen Ritter"*
Left: H. Müller's "p" on his MR137, painted by M. Kröbel; photo: J. Schwarz;
H. Schwahn's coll.; Right: Werner Kästner's *"Ritter aus Hessen"* design (TB 32)

According to Długosz, *"The banner of the **Knights of Meissen**, under which were 80 excellent lances* of brave Meissners, who at their own expense came to the aid of the Order with their own weapons. They did not want to fight under a strange Crusader flag, but only under their own lineage, family and House flag. ... This banner is 2 ½ ells long and 2 1/8 ells wide."*[36] The flag's cloth had four fields, of which the first and fourth were red and the second and third were blue; the sleeve was white. Werner Kästner produced a *"Banner der Ritter aus Meißen"* figure with a flag engraved only on one side (TB 34).

Knights of Meissen
Left: Golberg Go2896, painted & photo by R. Sanders.
Right: Werner Kästner's *Ritter aus Meißen* design (TB 34)

* The "lance" was the smallest tactical unit in many European armies of this period and it usually consisted of a knight, a squire and one or more fighters among whom was often a crossbowman, and theoretically all were mounted.

Among the Order's "guests" were some dukes, including the **Duke of Oels** about whom Długosz wrote: *"The banner of Conrad the Weissen, Duke of Oels in Silesia, which the duke led himself, were his own fighters from the Duchy of Breslau and from Silesia. But Duke Conrad was captured, his flag and all his possessions robbed, yet through the generosity of King Wladislaw of Poland he was set free.... This banner is 2¼ ells long and 1¾ ells wide. Let it be known that of all the dukes of Silesia and Poland none other than this Conrad von Oels and the Duke Casimir von Stolpe or Stettin were on the side of the Master and the Crusaders; to their ignominy and shame they were forced to maul and devastate their own fatherland. Both were however spared, like the prisoners, by King Wladislaw."*[37] The banner displayed a black Silesian eagle, on whose breast was a small white cross on a crescent in white, all in a yellow field. The eagle's tongue and talons are red. The cloth along the staff was decorated with a black bar. A "Banner of the Duke von Oels" figure with an engraved flag (TB 08) by Werner Kästner is still available. Oels is now Oleśnica, in south-western Poland.

Duke of Oels
Left: H. Müller's "r" on his MP23, painted by M. Kröbel; photo: J. Schwarz; coll.: H. Schwahn. Right: Werner Kästner's *Herzog von Oels* design (TB 08)

So-called "Knights from the Rhine and Germany"
Left: W. Kästner's *Vogtei Brattian* (TB 26) flag's blank side, painted & photo by R. Sanders. Right: Werner Kästner's *allgemeines Kampfbanner der Gastritter* design (TB 37)

According to Długosz: *"The banner of the **Knights of the Rhine and Germany**, led by the knight [illegible]. Under him were 60 lances of excellent Rhenish, German and Livonian knights, whom the Crusaders brought at their own expense and therefore flew their own and special banner from their lineage, nation and family. This banner is 2¾ ells long, 2½ ells wide, and has more white than the other colors."*[38] The flag's cloth was yellow in the upper field, white in the middle and red in the lower field; the sleeve was white. According to Heinz Schenzle it was actually a "common battle standard" (*"allgemeines Kampfbanner"*). Werner Kästner also produced an *"allgemeines Kampfbanner der Gastritter"* (common battle standard of the guest knights) figure with an engraved flag (TB 37), but the painted figure shown here used the unengraved side of the Kästner's *Vogtei Brattian* (TB 26) figure.

So-called "Banner of the Swiss Knights" (likelier 2nd Balga Commandery) H. Müller's MR57 with flag from Annex 1, painted & photo by R. Sanders

The so-called **"Banner of the Swiss Knights"** is definitely misidentified. The flag, which Długosz called *"The banner of the People and the Nation of the Swiss"* (*Banderium Gentis et nacionis Sweyczerorum*) is, according to almost all modern sources and experts, incorrectly identified. According to Wolfgang Büche, the publication in recent years of the Teutonic Order's accounting books proves that no knights from Switzerland took part in the battle. On the other hand, in Klein's assessment, the emblem, a wolf, on this flag is very similar to the flag of the Balga Commandery (see page 26). Based on this similarity, the hypothesis has been put forward that this flag was a second banner deployed by the Balga Commandery. Another possibility is that this division was commanded by a knight from the Frankish Fischborn family whose coat of arms also looked like the emblem shown on this flag.[39] This flag, with a white wolf with a black tongue in a red field was 1 3/8 ells long and 1 1/8 ells wide, that is, it was much smaller than the flag of the Balga Commandery. The author could not find any tin figures with a corresponding engraved flag.

So-called "Banner of the Swiss Knights" from *Banderia Prutenorum*
1 3/8 ells long and 1 1/8 ells wide

Banner of the Balga Commandery from *Banderia Prutenorum*
3 ells long and 1 7/8 ells wide

Duke of Stettin
W. Friedrich's engraved figure, painted & photo by H. Meißner
W. Kästner's engraved TB 09, painted & photo by R. Sanders

Another nobleman who was a "guest", who participated in the battle with a levy, was the Duke Kasimir of Stolpe or Stettin, about whom Długosz wrote: *"The banner of the **Duke Casimir of Stolpe or Stettin**, led by the duke himself and his people not without great dishonor and shame to his name, people and language, in that he was seduced by gold and who with anger in his family's guts, with all of his people capable of bearing arms, made war on his own fatherland and the Polish Kingdom from which he and his forefathers had their existence and lives. But after his people were exterminated in the defeat, he himself with his flag and some knights were captured and had to suffer the misery of prison for a long time yet was freed through the mercy of King Wladislaw of Poland. Only he and no other vassals* or blood relatives bore arms against the fatherland. ...*

This banner is 1 7/8 ells long and 1¾ ells wide, the tail iss 1 ¾ ells long, 1 ell wide and gets smaller and smaller at its end."[40] The flag had a white cloth with a tail on the upper portion and bore a red Pomeranian griffin. Its beak, claws and eyes were gold and it had black pupils and a black tongue, although the gold has turned to black over the ages; the sleeve was blue.

According to Ekdahl's research, Duke Casimir was not also the Duke of Stolpe so this banner should properly be referred to as that of simply the Duke of Stettin. Stettin is now Pomorze Szczecińskie, Poland.

The so-called "**Banner of the Knights and Mercenaries of Westphalia**," with its crossed red arrows in a white field, must again have been a "common archers" banner.[41] The Teutonic Order's accounting books again prove that no knights from Westphalia participated in the battle. According to Długosz, the banner was 2 3/8 ells long and two ells wide. Werner Kästner also produced a "*Banner der Ritter aus Westfalen - allgemeines Kampfbanner*" figure (TB 21) with a flag engraved on both sides.

So-called "*Banner der Westfälischen Ritter*" (more likely an archers' banner) Horst Tylinski's figure; painted & photo by R. Sanders

W. Kästner's "*allgemeines Kampfbanner der Ritter aus Westfalen*" design (TB 21)

Długosz recorded: "*The banner of the **Althaus Commandery**, was led by the Komtur of Althaus, Wilhelm Rippem, under him a few Teutonic Knights and local nobles but only mercenaries.*" According to Vossberg, the commander of Althaus who died at the Battle of Tannenberg was named Eberhard von Ippinburg but Długosz had garbled his name.[42] The flag's cloth consisted of four quarters, two white and two black. The size was not given in the *Banderia*, as it is one of the "*Rectobanners*". Werner Kästner also produced a "*Banner der Komturei Althaus*" figure with an engraved flag (TB 19). Althaus is now Starogród (Chełmiński), Poland.

Althaus Commandery - Left: Golberg Go2896, painted & photo: R. Sanders. Right: W. Kästner's design (TB 19)

*According to F. A. Vossberg, "The Pomeranian dukes were already German Imperial Princes from the 12th century; a vassalage relationship to Poland no longer existed in the 15th century."

According to Długosz: "*The banner of the **Commandery and Town of Balga** was led by the Komtur; under him were Teutonic Knights and some local knights and mercenaries. The banner is 3 ells long and 1 7/8 ells wide.*" The flag showed an erect wolf with black teeth, tongue and claws in a white field, above which was a black stripe. According to Vossberg, the *Komtur* of Balga at this time was Friedrich Graf von Zollern.[43] One should compare this flag with the so-called Banner of the Swiss Knights (page 24 above). In the aftermath of the Battle of Tannenberg, the Order Castle Balga was one of the few *Ordensburgs* castles that did not surrender to the Poles.[44] The town of Balga is now also called Weseloe, located in Russia's Kaliningrad Oblast. Werner Kästner's "*Banner der Komturei Balga*" figure (TB 16) has a considerably too large flag that is engraved on both sides.

Commandery and Town of Balga
W. Kästner's (TB 16) engraved,
Painted & photo by R. Sanders

Długosz wrote in the *Banderia*: "*The banner of the **Town of Bartenstein**, was led by the Steward (Vogt) of Bartenstein. Under him were some Teutonic Knights and men born in the Bartenstein District, who flew a flag with a kind of device the Poles call* Oxa *and the Bohemians call* Bradaczicza. *This banner is 2¾ ells long and 2¼ ells wide; the tail is 2¼ ells long, however only one hand wide and gets narrower and narrower to its end.*"[45] The flag had a black flag's cloth with a white hatchet or axe. On the top there was a narrow white stripe that continued as the tail; the sleeve was blue on top and black on the lower portion.

Town of Bartenstein
W. Kästner's (TB 28), engraved
Painted & photo: R. Sanders

The Town of Bartenstein (now called Barciany in Poland) was located in the Balga Commandery's district and its steward was subordinate to that *Komtur*. It has also been postulated by Ekdahl and others that the flag described in the *Banderia* as belonging to Heiligenbeil (see page 33) may have also been an adjunct to this banner, both being carried as part of the Balga Commander's subordinates. Even in medieval times the town's seal displayed crossed axes with long handles rather than the kind of axe on this banner.[46]

Bartenstein Town's seal
ca. 1915

In the *Banderia* Jan Długosz wrote: "*The banner of the* **Town of Brandenburg**, *which the Commander of Brandenburg, Marquart von Salzbach led. Under him were Teutonic Knights, men born there and Brandenburg burghers, as well as some mercenaries. This flag was granted to the city by the Margrave of Brandenburg described in documents when the Margrave raised a war party against the barbarians in Prussia. The aforementioned Marquard was however excellently captured, along with some Teutonic Knights, the flag mentioned, and with the knight Schumbork in the great battle, by Jan Długosz of Niedzielsko, a knight from the House of Zubrzaglowa alias Perstina or Wieniawa, my dear father and genitor. And when he [Marquard] was presented by my father to Alexander Vytautas* [or Vitold], *the Grand Duke of Lithuania, the latter was delighted at the sight and condemning him to death, on account of the fact that once at a meeting which both attended, he had within hearing of Vytautas called his mother a harlot and a dirty matron. Vytautas said to him 'Are you this Marquard?' who not forgetting his situation unreservedly and firmly answered 'Yes I am that one, who will calmly suffer the fate which befell yesterday, which can similarly befall you today or tomorrow.' Vytautas was even more insulted and embittered by this prideful speech, sentenced the speaker to death, although he had intended to spare him. And the Lithuanians and Ruthenians, on Vytautas's orders, took him to a cornfield and beheaded him there. When King Wladislaus of Poland learned this, he rebuked Duke Alexander strongly, assuring him that the victor had shamed himself by this treatment. Nothing is more praiseworthy, after such a victory than to have mercy on the unlucky ones. ... The banner is 2¼ ells long and the same wide.*"[47] This banner had a white cloth with the red Brandenburg eagle with a yellow beak, talons and gathering of feathers of the same color; the sleeve was blue, the staff light brown. This town called Brandenburg, in Polish named Pokarmin, is now called Uszakowe (in Russia's Kaliningrad Oblast), should not be confused with the principality of Brandenburg. According to Ekdahl, there was only a commandery and no town there at the time of the battle.[48] In the aftermath the Battle of Tannenberg, the Brandenburg castle was one of the few Order castles that did not surrender to the Poles.[49] Wolfgang Friedrich and Werner Kästner each produced a Banner of Brandenburg figure with an engraved flag (Kästner's TB 35).

Town of Brandenburg
Figure engraved, produced & photo by W. Friedrich, painted by J. Hensel; right: Werner Kästner's design (TB 35)

Banners from the 1410 Battle of Tannenberg, extract from the *Banderia Prutenorum*

In the *Banderia* it says: "*The banner of the **Stewardship (Vogtei) of Brathien and of Neustadt** was led by the Steward (Vogt) of Brathean, Johann von Reber; under him were some Teutonic Knights, some burghers of Neustadt and also mercenaries.*"[50] The flag was white with three yellow stag's antlers set together. The *Banderia* does not give any measurements for the flag. Brathien (Brathean or Brattian, now Bratian, PL) was a stewardship located in the Osterode Commandery but it was subordinate to the Grand Master in the Marienburg.

Brattien Stewardship and Neustadt
W. Kästner's "*Vogtei Brattian*" figure (TB 26) engraved on one side

Town of Braunsburg
W. Kästner's (TB 23) with flag engraved on one side only

Painted & photos: R. Sanders

For the flag of the **Town of Braunsberg** (today Braniewo, PL), the *Banderia* simply states "*The banner of the Town of Braunsberg. Under it were inhabitants and vassals, the burghers of Braunsberg as well as some mercenaries. This banner is 1¾ ells long and 1¼ ells wide.*"[51] The flag had a black cross in the upper white field and a white cross in the lower black field; the sleeve was black, the staff light brown. It is unusual for a coat of arms or banner to have a yellow heraldic device on a white field. Werner Kästner's "*Banner der Städte Elbing, Braunsberg und Danzig*" figure (TB 23), whose flag is engraved on one side, is suitable for the flags of these divisions from those three towns.

The *Banderia* supplemented the picture of the flag of the **Town of Kulm** (Culm/Chełmno) like this: "*The banner of the Town of Culm, which Nicolaus named Nyksz,* a Swabian, as the Culmer standard-bearer carried, whom the Prussian Grand Master later gave a death penalty for treason, whose leaders were Johannes von Orsichau and Conrad von Kropkow. ...Under him were the knights, the burghers of the Land and Town of Culm. ... This banner had a length of three ells and a hand's width, but only three ells wide; but the tail extended 3¼ ells; starting at only 7/8 ells and getting ever narrower.*"[52] The cloth was red decorated with two white rivers; a black Order's cross, but upside down,

Town of (*Stadt*) Kulm
Left: Golberg Go2895, Painted & photo by B. Grimm.
Right: Werner Kästner's design (TB 14)

Kulmerland - H. Müller's "b" on his MP 29
Painted by M. Kröbel; photo: J. Schwarz;
collection of H. Schwahn

was in the upper portion. There was a narrow black stripe along the entire length of the flag, including the tail.

A similar but not identical flag from **Kulmerland** was described thusly by Długosz: "*The army banner of the Land of Culm, was led by the Thorn Commander Johann, Grafen und Erben zu Seyn* (Count and Heir of Seyn); *under him were Teutonic Knights, men born in Thorn and Culm and burghers as well as mercenaries gathered by the Thorners. This banner is 2 3/8 ells long and 2 1/8 ells wide, the tail above was two ells long; it started one ell wide and got ever narrower.*"[53] The flag's cloth looked just like that of the flag described above from the Town of Kulm, but the cross was right side up and the tail differently marked. Werner Kästner also produced a figure with flag engraved on one side (TB 14) that can be used for both the Town of Kulm and Kulmerland.

Brattien

Braunsburg

Kulm

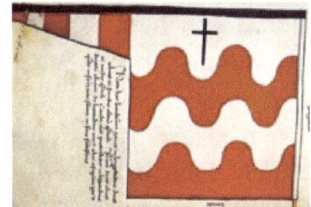
Kulmerland

---

* According to Vossberg, *Banderia*, page 38, Długosz had garbled the name; he was called Nicolaus von Renys, was head of the Lizard League (*Eidechsen-Bund*) and in 1411 was beheaded in Graudenz as a traitor to his land's lord. He is now considered as a scapegoat for the defeat at Tannenberg.

The Teutonic Order – diverse Hans Müller figures, painted & photo by Jan Sennebo (Helsingborg, Sweden)

Danzig Commandery
Werner Kästner's (TB 22)
Painted & photo by Bernt Grimm,

Commandery and Town of Danzig
Werner Kästner's (TB 23)
Painted & photo by R. Sanders

The first **Danzig** (now Gdańsk, PL) flag was described by Długosz as follows: "*The banner of the Danzig Commandery was led by the Danziger Komtur, Johann von Schönfeld. Under him were Teutonic Knights from the Danzig Order House and Convent, men born in Danzig and district noblemen, as well as foreign mercenaries. This banner is 2 5/8 ells long but the tail extends two ells in length and is ¼ ell wide getting narrower to its end. The banner is 1 7/8 ells wide.*"[54] A Danzig Commandery standard-bearer ("*Komturei Danzig Bannerträger*") figure (TB 22) was also produced by Werner Kästner.

Additionally Długosz recorded an identical but somewhat smaller flag: "*The second Banner of the Danzig Commandery and Castle, which the Vice-Commander of Danzig led with his Crusader brothers and paid mercenaries. Among which there were 70 lances of noble knights. This banner is 2½ ells long and 1 7/8 ells wide. The tail extends a length of two ells, but in width was ¼ ell and narrowed to its end.*"[55] The second banner was identical in form to the one above, but not as large.

The **third Danzig banner** was portrayed thus by Długosz: "*The banner of the Commandery and Town of Danzig was led by the mayor with burghers, men born there, mercenaries and people known as 'Schiffskinder'* [sailors, literally "ship's children"], *bold and courageous men who did not shrink from any kind of death; more appropriate for maritime than for land warfare. They increased the Danziger's banner to 100 lances and showed the image of their brave warriors in the fighting. The symbol on their flag was two white crosses in a red field. Casimir III, the King of Poland, who withdrew [Danzig] from Prussia, graced the Danzigers with a special privilege, a golden crown, as a sign of the change in rule, which they used thereafter.*[56] *This banner is 2 3/8 ells long and 1¾ ells wide.*"[57] It is likely that this flag was only carried by the city of Danzig and not the commandery. In the *Banderia Prutenorum* the red cloth showed two crosses under one another while the city seal from 1308 already shows the crown as mentioned above, perhaps just overlooked in Durink's illustration. Werner Kästner's "*Banner der Städte Elbing, Braunsberg und Danzig*" figure (TB 23), whose flag is engraved on one side, is suitable for the third Danzig flag.

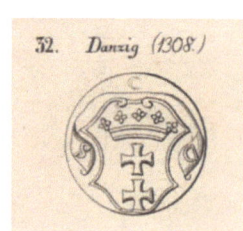

Danzig Komturei
1399 seal Vossberg
pg 215

Danzig 1308 seal
Vossberg plate XV

The levy and the flag of **Dirschau** (today Tczew, PL) was described in the *Banderia* so: *"The banner of the Stewardship and Town of Dirschau was led by the Dirschau Steward, Matthias Beberach, under him were Teutonic Knights and men from the district and Dirschau burghers. This banner is 2½ ells long and two ells and a hand's width wide."*[58] The cloth had four vertical alternating white and black stripes; the sleeve was ash gray. The painted figure by Werner Kästner shown here with an engraved flag (TB 20) was produced as the *"Banner der Komturei Tuchel"* while another figure in profile with a flag engraved on one side (TB 33) was intended as the standard-bearer of the Dirschau Commandery but was inaccurately engraved.

Dirschau Stewardship and Town
Left: Werner Kästner's (TB 20) painted & photo by R. Sanders; right: Werner Kästner's *"allgemeines Kampfbanner der Gastritter"* (TB 33) design

The *Banderia* states: *"The banner of the **Commandery and Town of Engelsburg** was led by the Komtur Burchard Wobeke. Under him were Teutonic Knights as well as some mercenaries."*[59] This Komtur died in the battle. Engelsburg was a commandery from 1278 until 1416; the town (now called Pokrzywno, PL), however, did not exist at the time of the battle according to Ekdahl's research.[60] The banner displays an angel with brown hair in a sky-blue gown in a red field; the flagstaff was light brown. The *Banderia* gave no information about the flag's size, a *"rectobanner"*. Hans Müller's MP29 figure, produced in the 1930s had an engraved emblem on the flag. Werner Kästner also produced an Engelsburg figure with a flag engraved on just one side (TB 24).

Engelsburg Commandery
Left: H. Müller's MP29, Painted & photo: Gianpaolo Bistulfi.
Right: Werner Kästner's (TB 24) design

Banners from the 1410 Battle of Tannenberg, extract from the *Banderia Prutenorum*

Elbing City and Commandery
Werner Kästner's TB 23 blank side

Elbing's City and Commandery 2nd banner - Werner Kästner's TB 23, engraved

Painted & photo: R. Sanders

In the *Banderia Prutenorum* there are three flags presented as coming from Elbing (today Elbląg, Poland). "*The banner of the Town and Commandery Elbing was led by the Komtur, Werner von Tettingen, under him the nobles and burghers of the Elbing District; as well as some Teutonic Knights and mercenaries.*"[61] Werner von Tettingen was the Senior Hospitaler (*Oberster Spittler*) as well as the Komtur of Elbing, who rescued himself from the defeat and remained in office until August 1412. The flag contained a white upper field with a red straight cross and a lower red field with a white cross; the sleeve was blue. The banner was 1¾ ells long and 1 1/8 ells wide. Werner Kästner's "*Banner der Städte Elbing, Braunsberg und Danzig*" figure (TB 23), whose flag is engraved on one side, is suitable for the 2nd Elbing banner and the engraved crosses can be modified for this flag.

Banner of the Burghers of Elbing
Hans Müller's MR320 with flag from Annex 1
Painted & photo: R. Sanders

"*The second Banner of the Town and Commandery of Elbing was led by the Elbing Vice-Komtur. Under him were Teutonic Knights, men born in the Elbing District as well as mercenaries. This banner is 2 7/8 ells long and 1¾ ells wide.*"[62] The flag contained a red Latin cross in the upper white field and in the lower red field a white Latin cross; the sleeve was blue. The flag of Werner Kästner's figure is engraved with these kinds of crosses.

The third banner from Elbing was described so by Długosz: "*The banner of the burghers of Elbing was led by the mayor. ... Under him were only Elbing burghers and some knights hired by him. ... This banner is three ells long and 1½ ells wide.*"[63] The flag contained a white straight cross in the upper red field and in the lower white field a red straight cross; the sleeve was blue. Obviously, this flag varied from the two banners from Elbing described above but the reversed colors correspond to Elbing's coat of arms as it appeared in the 19th century.

Banners from the 1410 Battle of Tannenberg, extract from the *Banderia Prutenorum*

Commandery and Town of Graudenz
Werner Kästner's (TB 15)
Painted & photo by R. Sanders

Town of Heiligenbeil
Werner Kästner's (TB 31)
Painted & photo by R. Sanders

The Graudenz levy's banner was described by Długosz as: "*The banner of the **Commandery and Town of Graudenz** was led by the Komtur of Graudenz, Wilhelm von Helfenstein. Under him were the knights and burghers living in and around Graudenz. The greater part of them used the head of a bison (żubr) as their emblem and for that reason displayed it on their flag. The commander, Wilhelm Helfenstein, was killed in the same battle along with many noble countrymen and mercenaries.*"[64] This white gonfanon flag had a black bison's head whose eyelids and nostrils were yellow and had an iron ring pulled through its nose. The *Banderia* gives no information about the flag's size, another "*rectobanner*". Graudenz was a commandery starting in 1250; today the town is called Grudziądz. The "*Banner der Komturei Graudenz*" (TB 15) figure by Werner Kästner, which is engraved on both sides, shows a cross over the bison's head although none appears in the *Banderia*. In Vossberg's study of Prussian medieval coins and seals, the seal for Graudenz bears neither the cross above nor the nose ring but has the bull's tongue protruding.

Długosz described the Banner of the **Town of Heiligenbeil** as follows: "*The Banner of the Town of Swiatha Siekierka, in German Heiligenbeil, was led by the Steward of Swiatha Siekierka, made up of his own Teutonic Knights and mercenaries. This banner is two ells long and equally wide.*"[65] The flag had a black field, a narrow white stripe above and a white axe in the middle of the field. Heiligenbeil is today the town of Mamonowo in the Russian Kaliningrad Oblast and still has an axe as part of its city coat of arms. According to Ekdahl's research, Heiligenbeil, which was located in the Balga Commandery's district had neither a steward nor an Order's castle, and the town's medieval seal showed crossed axes with long handles. This banner is very similar to that of Bartenstein also in the Balga Commandery's district. The provenance of this banner remains uncertain, but it could have been a second Balga Commandery banner carried by the House Commander.[66] Werner Kästner's Heiligenbeil figure (TB 31) has the emblem engraved on the flag and shield.

Graudenz seal (Vossberg, *Münzen*, plate VIII

Graudenz banner, (*Banderia Prutenorum*)

Heiligenbeil (Vossberg, *Münzen*, plate XV)

So-called Heiligenbeil town's banner, (*Banderia Prutenorum*)

So-called Bartenstein banner, (*Banderia Prutenorum*)

Regarding the banner of the city of **Königsberg**, the *Banderia* states: *"The banner of the Stadt [Town of] Königsberg was led by the Vice-Marshal and Vice-Commander of Königsberg by the name of [illegible] under him were some Teutonic Knights and mercenaries. The city flag displays as its image a white lion, which was granted to it by King John of Bohemia who at the time led an army against the barbarians... This banner is three ells long and 2 1/8 ells wide. The tail extends for 1¼ ells, is ¾ ells wide at the cloth and gets ever narrower."*[67] The flag had a narrow white field painted with an elongated Order's cross in black extending to the tail; the lower, primary, field was red with the white Bohemian lion rampant with a yellow crown and golden claws. Ekdahl considers it more likely that this banner was used by the Königsberg *Komtur* rather than by the city of Königsberg based on comparisons of seals and coins from that city (now named Kaliningrad).[68]

City of Königsberg
Werner Kästner's TB 18
Painted & photo by R. Sanders

Altstadt Königsberg
Werner Kästner's TB 36
Painted & photo by R. Sanders

A second Königsberg banner was described by Długosz thusly: *"The banner of the **Old Town (Altstadt) Königsberg**, alias Kroliowgrod, led by the Königsberger mayor, was composed of its own burghers and the people of the countryside as well as mercenaries. This banner had a length of two ells plus a hand's breadth and a width of 1¾ ells."*[69] The flag displayed a red royal crown in a white field above and below a white Order's cross in a red field; the sleeve was blue. Newer research assesses this flag as being from Kniephof, a city quarter of Königsberg (see Andrzej Klein, *Banderia Apud Grunwald II*, page 100).

The levy from **Leske** (now Leszken, PL) **or Lessen** (today Łasin, PL) is described in the *Banderia* thus: *"The banner of the Vogtei Lessen was led by Heinrich Kuszeczke,* the Steward of Lessen, named Zolawa in Polish. Under him were Teutonic Knights from the Marienburg Castle and some stewards and mayors of the towns and villages of Lessen. ... This banner is 2 ¾ ells long and 2 1/8 ells wide."* It had horizontal red, white and black stripes and a blue sleeve. According to Vossberg, regarding the name Heinrich Kuszeczke: *"From this name which Dlugoß badly garbled, one can

Stewardship of Lessen or Leske
Left: W. Scholtz 1051 with a flag from Annex 1,
Painted & photo by R. Sanders. Right: Werner Kästner's design (TB 27)

conclude that Conrad von Kunseck, the steward in office in Lessen since 1407, was present at the battle and died there because after this time there is no mention of him."[70] Ekdahl and others also assess this to be the banner of the Order's Steward of Lessen (now Łasin): he states that there was no commandery in Leske (now Leszken).[71] However, today's Lessen/Łasin town's flag shows these same colors in the same arrangement. Werner Kästner also produced a *"Banner der Vogtei und der Stadt Leske"* figure (TB 27) whose flag is only engraved on one side.

The image at the left shows a flag, which was described thus in the *Banderia*: *"The banner of the Town of **Groß-Holsten**, named Melsack in German. Led by the Steward of Holsten. Under him were Teutonic Knights from Holsten, men born there and burghers of the district, as well as mercenaries .... This banner is 3 1/8 ells long and two ells wide."*[72] The banner was composed of three horizontal fields – black, white and red; and the sleeve was in the same colors as the adjoining cloth. Because neither a stewardship nor a *"Pflegeamt"* (caretaker's office) existed in Melsack, the origin of this flag remains questionable. More recently this flag has been associated with the Town of

The so-called Banner of the Town of Groß-Holsten
Left: H. Tylinski figure 1184 with a flag from Annex 1, painted & photo by R. Sanders; right: Werner Kästner's design (TB 40)

Allenstein (Olsztyn, PL). Werner Kästner also produced a *"Banner der Stadt Allenstein"* figure with an engraved flag (TB 40), but the flag is too large.

According to Długosz: *"The Banner of the Komturei **Nessau** (Nieszawa) was led by the Komtur Gotfried von Haßfeld and under him were Teutonic Knights and mercenaries."*[73] This commander was killed at the Battle of Tannenberg. The flag had three equal-sized vertical fields, black-white-black. The *Banderia* gave no information about the size of the cloth, yet another *"rectobanner"*.

Left: Banners from the 1410 Battle of Tannenberg, extract from the *Banderia Prutenorum*

Nessau Commandery
Wohlmann's DOR5 with a flag from Annex 1 Painted & photo: R. Sanders

The so-called "*Banner of the Commandery, the Castle and the Town of Gnyew or Mewe*" **(image to left)** was definitely misidentified by Długosz.[74] It most probably was a common standard for archers. As mentioned above, it can be concluded that the flag of the Mewe Commandery was a version of the current coat of arms of the town, Gniew, Poland, as in a seagull (German *Möwe*). The seal of the Komtur and House Commander of Mewe which also displayed a seagull was such a "canting" coat of arms. From the *Banderia*, we can only assert that the red flag showed the crossed pointed and blunt arrows in white and that it was 2 1/8 ells long and 1¾ ells wide.

So-called Mewe Commandery Banner
Stoll's RTP 32b with flag from Annex 1
Painted & photo: R. Sanders

Concept for Mewe Commandery banner
Painted & photo: Wolfgang Büche

It has also been suggested that this flag and the very similarly designed flag attributed to the Swabian Knights (who were not present at the battle) were from the same unit. Or perhaps the Order had the flags produced to use with the consolidated archers and crossbowmen drawn from the various "lances".

The author was unable to find any flat tin figures with a correspondingly engraved flag.

Mewe town's 14th C. seal (Vossberg, *Münzen*, p. 215

So-called Banner of Mewe - 2 1/8 ells long and 1¾ ells wide

So-called Swabian Knights Banner - 2 ells x 2 ells

Schwetz or Ortelsburg Commandery

Schwetz or Ortelsburg Commandery
Werner Kästner's TB 38 design from the Schmalkalder ZF catalog

There are still questions about the red and white, diagonally colored flag **shown above**. In the *Banderia*'s page (*folia*) 18a, it states: "*The banner of the **Commandery and Town of Scythno [Schwetz]** was led by Komtur Grafen Albert von Eczbor zu Ortelsburg or Scynthno. Under him stood Teutonic Knights and resident vassals belonging to the **Ortelsburg Commandery**. ...This banner is 2 ¾ ells long and 2 ¼ ells wide.*"[75] According to Vossberg, this commander died in the battle. The flag, seen from the right side, was divided diagonally with a red upper and white lower field; the sleeve was blue. This flag looks almost the same as that of the Senior Draper (*Oberster Trapier*) (see page 20 above). It should be noted that after the Battle of Tannenberg, the Ortelsburg castle at Schwetz (Szczytno, PL) was one of the few Order castles that did not surrender to the Poles.[76] Along with Kästner's "*Oberster Trapier*" figure (TB 06 on page 20) mentioned above, he also produced a figure with a similarly engraved flag designated as "*Komturei Schwetz*" (TB 38).

In the *Banderia* it states: *"The banner of the **Town and Commandery Osterode** was led by the Osterode Komtur Penczenbrun, whose standard-bearer was Peregin Vogel, the Osterode ensign. Under him were both Teutonic Knights and resident vassals of the Osterode Commandery. ... This banner is three ells long and 2 1/8 ells wide."*[77] The flag's cloth had four fields, of which the first and fourth were white and the second and third were red, while the sleeve was white. According to Vossberg, in the Kraków Manuscript the *komtur*'s name appears as "Pentzenheyn". It was, however, Gamrath von Pinzenau, who died at the Battle of Tannenberg. Werner Kästner also produced a *"Banner der Komturei Osterode"* figure (TB 29) with a flag engraved on just one side. Today the town of Osterode is called Ostróda (PL).

Osterode Town and Commandery
Left: Golberg Go2897, painted & photo: R. Sanders; right: Kästner's design (TB 29)

So-called Banner of the Town & Commandery of Ragnit
Werner Kästner's (TB 39), engraved
Painted & photo by R. Sanders

According to the *Banderia*, *"The banner of the Town and the **Commandery of Ragnit** was led by the Komtur of Ragnit, Count Friedrich von Zollen, under him were the Ragnit Convent's Teutonic Knights and the vassals born within the Commandery from the Ragnit District. ...This banner is 2½ ells long and 2¼ ells wide."* It has already been mentioned that according to Vossberg this flag was mistakenly attributed to the Bishop of Sambia (*Samland*). He wrote further that *"It has already been mentioned that the flag ([in my] plate II, No. 9) that was mistakenly attributed by Dlugoß [sic.], completely corresponds to this one, and that the two flags with the three hats probably belonged to Ragnit. Because we indeed know from an existing seal, with St. Peter and St. Paul standing, that the commander of Ragnit around this time was named Helfrich von Drahe, as opposed to the Count Graf Friedrich von Zollern who was present as the*

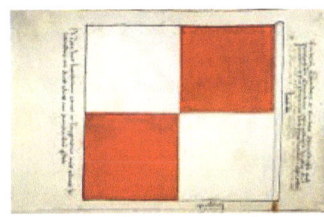
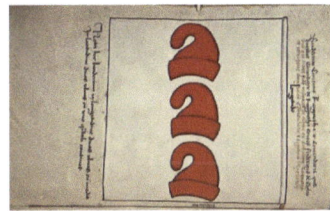
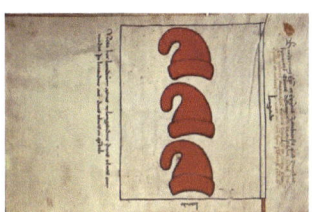

Osterode Commandery
(*Babderia Prutenorum*

Center: Town and the Commandery of Ragnit (folio 25): right: So-called "Bishopric and Bishop of Samland" (Sambia) (folio 9) – both 2 ½ x 2 ¼ ells (*Banderia Prutenorum*)

Commander of Balga at the Tannenberg battle, so the interpretation of these flags must remain uncertain."[78] These two flags' same appearance and size indicate that they were carried by the same levy. Ragnit (now Neman in the Russian Kaliningrad Oblast) was founded as a commandery in 1289. After the Battle of Tannenberg, the Ragnit Castle was one of the few *Ordensburg* castles that did not surrender to the Poles.[79]

Roggenhausen Stewardship & Town
Left: H. Müller's "m" on his MR137, painted by M. Kröbel; photo: J. Schwarz;
collection of H. Schwahn; right: Werner Kästner's figure (TB 41)

According to the *Banderia Prutenorum*, "*The banner of the **Stewardship and Town of Roggenhausen**, was led by the Teutonic Knight and Steward Friedrich von Wenden. Under him were the knights of the Roggenhausen District from the Order's House and the Dolywa family, for whose decoration and honor the stewardship was clothed with such signs, which shows that it belongs to the Kingdom of Poland and was populated and possessed by the Poles.*"[80] According to Vossberg, the Steward of Roggenhausen was Friedrich von Wenden who died at Tannenberg, and according to Ekdahl, there was no town at Roggenhausen at that time[81] but otherwise this description is mostly correct. Roggenhausen, now Rogóźno, Poland, is located in the Warmia region (*Ermland*). The flag displayed a red diagonal bar decorated with three white roses in a white field. The manuscript gives no information about the flag's size, a "*rectobanner*". Werner Kästner produced a "*Banner der Stadt von Roggenhausen*" figure at a trot and with an engraved flag (TB 41).

Banner of the Roggenhausen Stewardship – Figures by the former Golberg (now Dangelmaier).
Painted & phot0 by R. Sanders

Commandery and Town of Schlochau
(likelier Rheden Commandery)
Werner Kästner's (TB 13)
Painted & photo by R. Sanders

The flag, which Długosz identified in the *Banderia* as "*The banner of the **Commandery and Town of Schlochau**"*, was almost certainly borne by a different division. Długosz's designation must be incorrect because the Schlochau Commandery's seal displayed a prone bull with an Order's cross.[82] However, the seal of the *Komtur* of **Rheden or Rehden** pictures the "lamb of God".[83] The flag's red upper field, with a red tail, displayed God's lamb in white holding a cross-flag and with the lamb's blood spraying from its heart into a chalice standing in front of it. The lower, narrower field was white like the sleeve. The banner was 2¼ ells long, 1¾ ells wide; the tail extended for two ells, starting with a width of ¾ ell and tapering to a point.[84] Rheden or Rehden (Polish Radzyń Chełmiński) was a commandery starting in 1251. After the Battle of Tannenberg, Rehden Castle was another of the few Order castles that did not surrender to the Poles.[85] Note, the differences in the lamb's halo and the shape of the flag between the flag shown below for the Rheden Commandery and that for the Bishopric of Warmia on page 18. It is highly conceivable or even likely that these two similar banners are simply two banners from either the Rheden Commandery or the Warmian Bishopric rather than being from two distinct entities and distant locations. Rehden/Radzyń Chełmiński is 120 km from Braunsberg/Braniewo the Warmian Bishop's initial seat and 70 km from Lautenburg/Lidzbark where the seat was in 1400.

The **Schlochau Commandery**, established in 1323, almost certainly had a different banner. Vossberg calls into question this banner as actually being from Schlochau (now Człuchów, PL) based on a *Komtur's* seal showing a lying bull with an Order's cross, and the town's seal of a bull's head with it turned and with an open mouth.[86] According to a website about the heraldry of the town of Człuchów, the bull is shown in white on either a blue or red field.[87] Another source, Engel, who is mentioned in both the town's website and Ekdahl's work, contends that the animal is a goat rather than a bull.[88] Of the five possible reconstructions of the banner shown below, the last two with the halved fields seem unlikely based on the patterns of other commandery banners. In Kästner's "Schlochau" figure (TB 13), the upper field is too small and the lower field too large.

So-called banner of Schlochau Commandery (*Banderia Prutenorum*)

So-called banner of the Bishopric of Warmia (*Banderia Prutenorum*)

Schlochau – the oldest known seal of the Komtur of Schlochau from the 15th century; early coat of arms[89]

Schlochau Commandery – reconstruction of possible banners from Matwijewicz's article
(https://ugczluchow.pl/files/file/Grzesiek/Geneza.pdf)

Regarding the **Schönsee Commandery**, besides the depiction of a white flag with two red fish, the *Banderia* only tells us *"The banner of the Commandery and Town of Schönsee was led by the Komtur of Schönsee or Kowalewo, Nicolosch Wylcz; under him were Teutonic Knights and some mercenaries and a few knights."* F. A. Vossberg wrote in addition that *"Nicolaus von Vilz (Filz), the Komtur of Schönsee since 1399, fell in the Battle of Tannenberg."*[90] Schönsee is now Kowalewo, Poland.

Commandery and Town of Schönsee
Werner Kästner's (TB 17)
Painted & photo by R. Sanders

Town & Commandery of Strasburg
Left: H. Müller's "l" on his MR167, painted by M. Kröbel,
photo: J. Schwarz; coll: H. Schwahn. Right: Werner Kästner's design (TB 25)

Regarding the Banner of the **Town and Commandery of Strasburg** we read: *"The banner of the Town and Commandery Brodnica was led by the Baldewin Stal, Komtur of Brodnica or Strasburg; under him were Teutonic Knights and district warriors, as well as some mercenaries. This banner is 1 ¾ ells long and 1 ½ ells wide."*[91] The flag displayed a red stag in a white field; the sleeve was blue. According to Vossberg, this Komtur died at Tannenberg. Strasburg, today Brodnica, PL, was the seat of a commandery starting in 1337. Shortly after Tannenberg, the castle surrendered to the Poles and it was granted to one of the Polish knights.[92] Kästner produced a *"Komturei Strasburg"* standard-bearer with an engraved flag (TB 25).

Długosz recorded in the *Banderia Prutenorum*: *"The banner of the **Town of Thorn** [Toruń, Poland] was led by the Thorner mayor, with residents, mercenaries, many foreign warriors and mercenaries recruited by Thorners for pay, which made up 80 lance-bearers in number. ... This banner is 2 1/8 ells long and 1 7/8 ells wide. The tail extended to a length of 1¾ ells was 1¾ ells wide starting at the top and getting narrower to the end"*[93] The flag's cloth along with the tail were white and it bore a red gate with three

Town of Thorn
Werner Kästner's (TB 30), engraved   Werner Scholtz's 1033
Painted & photos by R. Sanders

towers, the window and door openings were black, the doors yellow with black hinges, and the portcullis white. According to Vossberg, Thorn's mayor, "*Albrecht Rothe († 1421), was highly respected and used often by the Grand Master as an emissary, for example to the King of Poland in 1414.*"

Commandery and Town of Tuchel
Golberg's Go2901
Painted & photo by R. Sanders

"*The banner of the **Commandery and Town of Tuchel** was led by the Teutonic Knight Heinrich as the Komtur. Under him were Teutonic Knights as well as district mercenaries. This Heinrich, a Frank from Germany, demonstrated from the beginning until the end of the war, such conceit, that he had two unsheathed swords carried before him, no other than if he was already the victor and glorious, triumphant one and as if any success in war would be dependent on him. This was onerous to the Grand Master Ulrich and his commanders.*" According to Vossberg, the *Komtur* of Tuchel was Heinrich von Schwelborn and he was taken prisoner in the battle and beheaded. The commandery's flag had two vertical fields, the white one next to the staff and the red one to the outside. Vossberg wrote further "*Because the coat of arms on this Tuchel Commandery banner, as we are acquainted with the same device from a seal from the same, it must be concluded that Długosz did not arbitrarily designate this banner.*"[94] The Commandery at Tuchel (today Tuchola, PL) was founded in the eastern part of the Schlochau Commander's district in 1330 but until 1384 Tuchel was administered by stewards. Werner Kästner produced a figure as "Tuchel" (TB 20); it is shown further above (see page 31) painted as "Dirschau", because the flag is incorrectly engraved for Tuchel.

Schönsee Commandery

Strasburg Commandery
Banderia Prutenorum

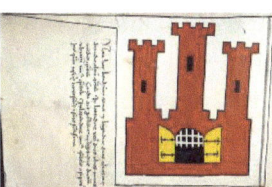
Thorn Town

Tuchel Commandery

The Teutonic Order – diverse Hans Müller figures. Painted & photo by Gernot Schlager (Leonding, Austria)

Possible Order's Banners Not Captured at Tannenberg/Grunwald

It is highly unlikely that the banners captured by the Polish-Lithuanian forces at the Battle of Tannenberg/Grunwald constituted the entirety of those borne by the Teutonic Order and its allies. Therefore, the following possibilities are presented for consideration as units and their possible standards that could have been present at the battle. Based upon the symbology in the banners shown in the *Banderia Prutenorum*, it would seem that the banners for the following units could have corresponded to those used on the seals or coins from their towns or districts or could simply have been geometric designs like the banners of the Dirschau or Gdańsk commanderies.

Barten (Barciany, in the Warmia-Mazuria region of PL) Trustee's Office. Its Order Castle, built starting in 1325, was initially planned as a commandery castle, but it became a trustee's office instead in 1354 (in Brandenburg Commandery). It was a temporary seat of a commander. Conceivably, its flag could have been red bearing an axe device in white, based on the traditional coat of arms of the town of Barten.

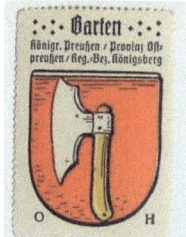

Barten coat of arms from a ca. 1915 card

Birgelow Comth 1260 seal Vossberg *Münzen*, plate XIV No. 16

Birgelau, also called Birglau (now Bierzgłowo, PL; English: Birgelow) was the seat of a Teutonic Knights commander from 1270 to 1415. Its castle was erected in 1232 and expanded until 1305. Further work and expansion of the castle took place in stages from the end of the 13th century to the beginning of the 14th century, although in 1415 the castle lost the function of the convent's seat,

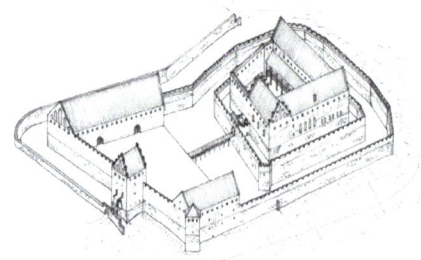

Reconstruction of Bierzgłowski Castle (medievalheritage.eu)

transforming into the seat of the Teutonic trustee (MHG *pfleger*). The flag could have featured some version of the device shown in the coat of arms as it appeared in 1260.

Bischöflich Papau (Papowo Biskupie, PL). The Papau Commandery's flag was not captured at Tannenberg although it is likely that its members participated in the battle.[95] This location was captured by the Order as early as 1232 and it became a commandery in 1287. Interestingly, it is the only Order Castle that is completely constructed of fieldstone rather than being at least partially made of bricks and it served to secure the

Modern coat of arms of Papowo Biskupie

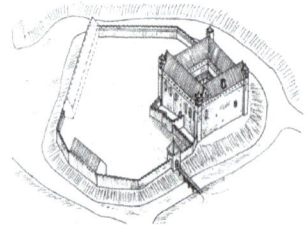

reconstruction of the castle from north-west according to B.Wasik (medievalheritage.eu)

route between Thorn (Toruń) and Kulm. After the defeat at Tannenburg, the castle fell into Polish hands but was soon taken back by the Order. With regard to its flag, it is certainly possible that it used that of Kulm or Kulmerland (see pages 28-29) or some variation with an eagle as displayed in its modern coat of arms above.

Bütow (Bytów, PL). Bütow was the location of a trustee's office in the Marienburg Administrative District. Before 1412, it was subordinate to the Danzig Commandery, which was confirmed as having participated at the battle. Bütow's castle was built between 1399 and 1405. Its banner could have been some form similar to that of the current Bytów Province flag that shows a black griffin with red talons and beak on a yellow field on the left half and two rows of five six-pointed yellow

Bytów Province flag

Bütow - town's 1329 coat of arms, Vossberg, *Münzen*

stars on a sky-blue field on the right side. Conversely, it might simply have used the Danzig Commandery's flag as shown on page 30, or some variation of the castles in the coat of arms from 1329 as shown here.

Gollub's traditional coat of arms
(https://pl.wikipedia.org/wiki/Herb_Golubia-Dobrzynia)

Gollub Komturei image by Jaume Ollé
(http://areciboweb.50megs.com/fotw/flags/de_to-gl.html)

Gollub (Golub, PL) became a commandery starting 1293 and lay within the Kulmerland area. This commandery's flag was not captured at Tannenberg although it is highly likely that this commandery participated in the battle.[96] Gollub's traditional coat of arms displays a Teutonic Knight flanked on both his sides by a bird (dove?) perched on a tree trunk without branches or leaves. It is difficult to hypothesize the appearance of the banner based on the coat of arms; however, the Flags of the World (FOTW) website suggests the banner would or could have been a white dove holding a green branch in its yellow beak, all on a red field.

Insterburg (Russian: Chernyakhovsk, RUS, Kaliningrad Oblast). The Insterburg, a brick castle, was a trustee's office (*Pflegeamt*) starting in 1347. It became a commandery and administrative center for the Nadrauen region, within the Königsberg Administration District of the Senior Marshal's Office. The banner probably displayed a black bear, walking, on grass, in a white field, extrapolated from the town's current coat of arms.

Chernyakhovsk (German Insterburg) modern coat of arms

Memel 1252 seal Vossberg *Münzen*, plate XIII No. 12

Klaipeda coat of arms
en.wikipedia.org

Memel. The Commandery at Memel (in Couronia, now Klaipėda, LIT) was founded by the Livonian *Landmeister* as part of the Order of the Sword Brothers (*Schwertbrüderorden*) and was established as a commandery in 1253. It was passed to the Teutonic Order State in 1328. The Order castle belonged to the Königsberg Administration District of the Senior Marshal's Office. The castle was split between the Order and the Curonian Bishop in 1352. Starting in 1393 it also administered the Memel Commandery's territory. After the Battle of Tannenberg, the Memel Castle was one of the *Ordensburgs* (Order castles) that did not surrender to the Poles.[97] Memel's seal, as early as 1252, featured the image of a castle keep flanked by wooden towers on either side and a ship below the fortifications. That device in yellow on a red field remains today in Klaipėda's flag and coat of arms, so it might be concluded that the commandery's banner would have featured such a device and color scheme.

Coat of arms at Przezmark Castle, PL

Preussisch Mark (Pruschenmarkt or Preuschmark) - Przezmark, PL. There is documentation of the Order having established a castle at Preussisch Mark, in Warmia/Ermland as early as 1274. After the Battle of Tannenberg, the castle was surrendered to the victorious Poles.[98] It has not been possible to definitively identify a coat of arms or flag associated with this entity; however, an added device showing a lion rampant with a diamond in each forepaw that appears on the restored Przezmark Zamek (castle) may give a possible option for extrapolating the banner.

Rhein, (Ryn, PL). This commandery was established in 1394 near the Wilderness (*Wildnis*) which was also designated as "Rhein Area". The Commander's seat changed at times between Rhein and Rastenburg. When Rhein was still part of Prussia, its coat of arms displayed a stag resting in front of a tree and that device remains the symbol on the flag of Ryn, Poland. Thus it is probable that the commandery's standard would have also displayed the stag and tree device.

Rhein - Ryn Kaffe Hag album 1915

Concept of Rhein Commandery banner based on current Ryn flag

Soldau wiki.genealogy.net/ Soldau

Soldau 1349 seal Vossberg *Münzen*, plate XVII No. 64

Soldau (Działdowo, PL). Soldau was the seat of a commandery starting in 1350. Its banner might have appeared as the red and white checkered pattern that appeared to the left and right of the queen displayed in Soldau's seal from 1349 and its later coat of arms. The town's current flag displays that kind of checkered shield in the upper corner of a *per fess* (halved) yellow over sky-blue field. So it is very likely that the Soldau Commandery's banner could have simply been a red and white checkered flag in five rows or that coat of arms shield on a different colored field.

Wenzlau Trustee's Office. From 1289 to 1326, Wenzlau (Unisław, PL) was the seat of a commander; after 1326 it was a *Pflegeamt* (trustee´s office). A flag from this *pflegeamt* was not captured at Tannenberg although it is likely that it participated in the battle.[99] Given the coat of arms of the town of Unisław, it is possible that the Trustee's office could have had a flag with the same device, possibly on a white rather than a red field, as many Teutonic Order flags were white rather than red (favored by the Poles).

Coat of arms of Unisław, Poland

The Livonian Banners Captured in 1431

In the *Banderia Prutenorum* four flags were depicted that were not present at the Battle of Tannenberg because the Livonian forces did not participate in that campaign by the Teutonic Order. Nonetheless, these flags are presented here because they were included in the *Banderia Prutenorum*.

The Teutonic Order in Livonia
K. & W. Wohlmann's DOR3 with a flag from Annex 1. Painted & photos by R. Sanders

Długosz wrote: "**The Banner of the Master of the Livonian Order** of Crusaders, which Theodrich Croe, the Livonian Marshal, led in the 1431 engagement at the village of Dambky next to the Verse [Verissa] River at Nakel. He was told about the area by Jodocus von Hohenkirchen, [the new] Commander of Tuchel. ... This flag had two devices, namely a picture of the Holy Virgin Mary. And the other side had the image of Saint Mauricius. ... This flag is also 2¾ (sic) ells long and the same width" (per Vossberg's translation, page 60). However, in the Latin text by Długosz it is 1 ¾ ells long and wide.

Długosz wrote further: "*The banner of the Livonians was led by Theodric Croe, the Marshal of Livonia, in the engagement in the year 1431. Under this flag, with its two signs and pictures, namely of the Virgin Mary and St. Mauricius, as well as two black crosses, were 300 lances of chosen knights, except for the foot soldiers which were fully practiced in warfare with flashing weapons. He, Th. Croe, the Marshal of Livonia, personally with many other commanders and his and foreign knights were captured and held for a long time in close captivity in the St. Bernard Tower of Kraków Castle but finally freed from there by the mercy of Wladislaw II, the King of Poland...*" It then states "*This banner is also 1¾ ells long and equally as wide.*"[100]

 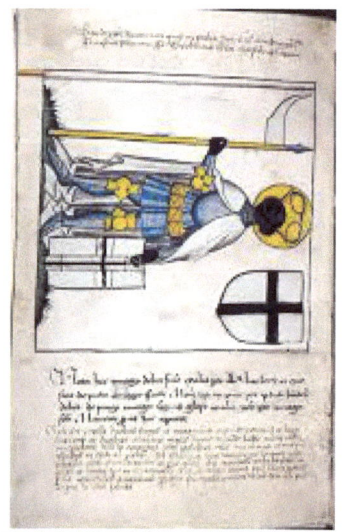

The Banner of the Master of the Livonian Order extracted from the *Banderia Prutenorum* and another source

The "Land Master" (*Landmeister*), the title of the head of the Livonian branch of the Order, had his seat at the Wenden Castle (Cēsis, Estonia).

Ascheraden and Dünaburg Commandery
Left: H. Müller's "j" on his MR129, painted.: F. Bähr; coll: H. Schwahn; Right: Extract from the *Banderia Prutenorum*

Describing the banner of the **Ascheraden and Dünaburg Commandery**, Długosz wrote in the *Banderia* "*The Livonian Banner captured in the fighting on 5 February 1431, before the celebration of the crucifixion, by the Polish Army at the village of Dambky by the Verse (Wyrzscha) River next to the town of Nakel, where Joh. Jarogniowski, Barthossius Vizemburgk and Dobrogost Kolynski with the help of the Steward led the Lithuanian army consisting of 700 knights headed by seven commanders inflicted a great, crushing defeat. The two leaders of this banner were Livonians, one namely Szwor, the Komtur of Ascheraden, the other Walter de Lo, the Komtur of Dünaburg; both of whom were killed in this engagement. But the banner [unit] mentioned had a covering force of only 200 lancers, except for foot soldiers, who were all miserably killed by the attackers (foot troops) in the engagement itself or in the woods and bushes where they fled to hide. ... This banner is 2 3/8 ells long and 2 ¼ ells wide. This banner has 3 tails as shown here. Each of these tails extends for one ell and is ¼ ell wide and then gets increasingly narrow until the end.*"[101] The flag had a black cloth and tails, with two six-pointed white stars arranged vertically. Ascheraden (now Aizkraukle, Latvia) was probably the seat of the Livonian Master from 1305-1387.

Fellin Commandery – *Komtur* of Livonia
K. & W. Wohlmann's DOR 5 with flag from Annex 1
Painted & photo: R. Sanders

The flag that is designated from the Banner of the **Fellin Commandery** and its unit are described in the *Banderia* as: "*Commander of Livonia - The Livonian banner which was taken in the engagement, was under the commander Joh. Jarognyowski from the House [and Family] of Bartoz Wizemburk of the House [and Family] of Dobrogost Kolinski, at the village of Dambky by the Verse [Wyrzcha] River next to Nakel, by just the Polish foot soldiers from the Livonian Army [on February 5] before the celebration of the crucifixion. The leaders were the two Livonian commanders, namely Walter Gyelze, the Komtur of Fellin, and Walter von Kerzdorf, the Komtur of Curow, both taken prisoner and thrown into the St. Bernard Tower's dungeon in the Kraków Castle and who died there. The Banner is two ells and one hand's breadth long and two ells wide; has also 3 tails, each of which extends for a length of one ell and starts as ¼ ell wide and narrows to a point.*"[102] The upper field was white with a black tail, the middle field black with a white tail and the bottom field was white with a black tail. Fellin (Viljandi, Latvia) was the seat of the Livonian treasurer (*Tressler*).

The last plate in the *Banderia Prutenorum* shows the flag of the **Stewardship of Kokenhausen**, which was described by Długosz thusly: "*The Livonian banner which was captured in the same engagement under the same leaders as before. The leader of it was Georg Eutzlebe (Kuthzlebe), the Steward of Kokenhausen, and he had 100* *lancers all of whom were killed by the peasantry. The peasantry's anger and embitterment knew no mercy after the Livonian Army had of late destroyed and burned down their villages' houses and food. But there were, in the Livonian Army, people and warriors from Pomerania, Culm and Prussian Livonia, from Courland and a great number of peasants. The Prussian Master, Paul von Kuszdorff (sic Nußdorf), had the Prussian commanders bring help to this army."* The flag was identical in appearance to that of Fellin, but a little smaller; it was 1 3/8 ells long and wide and *"had 3 tails, each of these extended to ¾ ell and were ¼ ell wide and narrowed to a point."*[103] Kokenhausen (today Koknese, Latvia) was also called Kokenhusen.

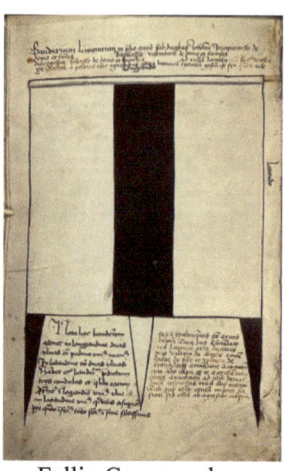

Fellin Commandery (*Banderia Prutenorum*)

Stewardship of Kokenhausen (*Banderia Prutenorum*)

The Battle of Tannenberg from the *Berner Chronik* by Diebold Schilling, ca. 1484

BANNERS OF THE POLISH-LITHUANIAN ARMY AT TANNENBERG

Left to right: Banner of the Polish Kalisz District (W. Friedrich), Banner of Żmudź (H. J. Stoll), Banner of Smolensk (K. & W. Wohlmann); Banner of the Tartar forces (W. Friedrich) - Painted & photo by R. Sanders

The Polish-Lithuanian Army and its Disposition at Tannenberg

The Polish Army consisted of two nearly equal-sized contingents: the Polish wing and the Lithuanian-Ruthenian, to which the Tartars belonged. Various military historians estimate that around 20,000 men served in the 51 "banners" (divisions or units under a given banner) of the Polish Crown and about 10,000 under the 40 banners of the Lithuanian-Ruthenian forces. Infantry, primarily peasants and craftsmen with primitive weapons, was positioned behind the Polish Army's formation and it is hard to estimate their numbers but they would play a key role in the battle. The combined Polish-Lithuanian army was numerically superior to the Teutonic Order's army but – primarily regarding the Lithuanians – more poorly armed.

Günter W. Fricke wrote "*According to the Polish Army Constitution, the barons, i.e., voivodes, castellans and other great feudal lords, who all possessed corresponding land holdings, had to provide their own trained units. These barons mostly personally commanded their troops. Some, however, tried to evade their duties as vassals. For example, the Magnate Jan Jemczikowicz sent mercenaries [Annex 7] (plate 1, banner 7). The members of the petty nobility either joined a baron or fought under their region's banner. They would have provided the bulk of the warriors in the knights' brotherhoods. Commoner fighters wore the coats of arms of their lord or that of their land on their shields. Besides the banners of the regions, there were the royal banners under which Poland's bravest knights (plate 1, banners 1 to 5) and the guests as well as also the King's mercenaries (plate 1, banners 6 to 8) fought. The Small Royal Banner was the standard flown by the King's bodyguard of the 60 lances.*"[104]

The mass of the Polish contingent was distributed under the 56 banners of the various voivodships and principalities. The supreme commander and commandant of the Polish Contingent was King Wladislaw II Jagiello. The direct command over the Polish contingent was held by Zbigniew z (of) Brzezie, Marshal of the Polish Kingdom, who also led his own 34th Banner. Under his command were also additional noblemen:

    Zyndram z Maszkowic, the head of the Great Camp of the Crown and the Commander of the Banners of Kraków;
    Krystyn z Ostrów, the Castellan of Kraków (26th Banner);
    Jan z Tarnów, the Voivode of Kraków (27th Banner);

Sędziwój z Ostroróg, the Voivode of Poznań (28th Banner);
Mikołaj z Michałów, the Voivode of Sandomierz (29th Banner);
Mikołaj Trąba z Wiślicz (44th Banner);
Pjotr Szafraniec z Pieskowa Skała, the Chamberlain of Kraków and Marshal of the Court (35th Banner).

Regarding the "barons", i.e., the King's officials, Günter W. Fricke wrote how they were differentiated as castellans and voivodes: *"In Poland in the early Middle Ages, the castellan was an official of the prince in whose name he undertook military and judicial administrative functions. Later he became a member of the Senate and commander of the general levy in the prince's region (pospolite ruszenie). At this time, the Voivode was the most senior state official who administered the King's region. As the deputy to the King, he also functioned as the army's commander (wojsko = army, dowódz = to lead)."*[105]

The Lithuanian contingent and its banners on the right wing were under the command of Grand Duke Vytautas. Among Vytautas's allied forces were approximately 3,000 Tartars under Khan Jalal el-Din, who with their archers would play a significant role in the battle. Prince Simeon Lingwen (Lingven), the brother of King Jagiello, received command over the three Ruthenian banners from Kiev and Smolensk.

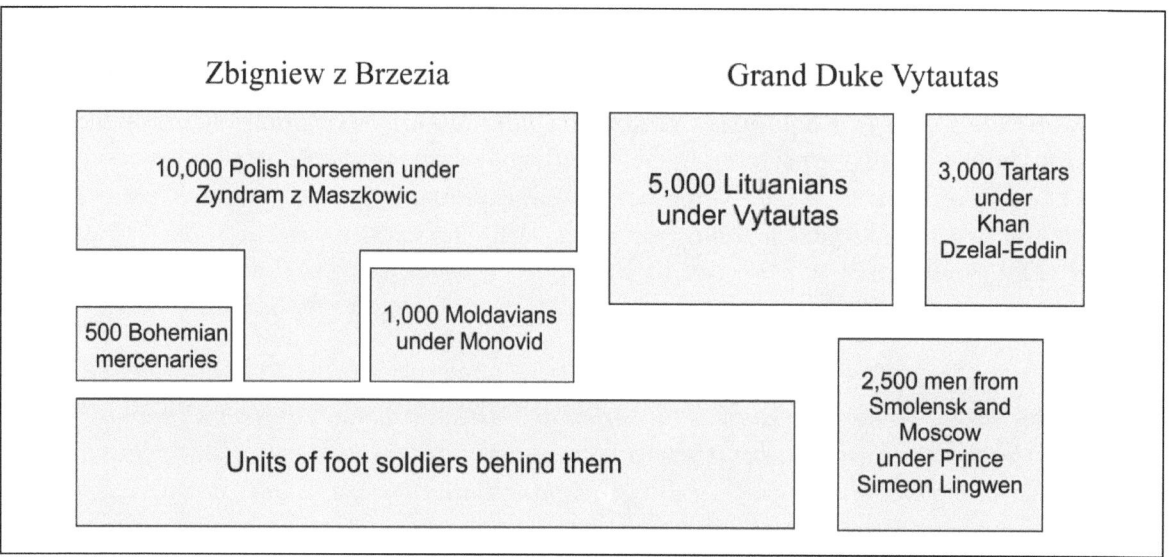

Dispositions of the Polish Army (after Heinz Schenzle's *Schlacht bei Tannenberg 1410 Kampf um die Marienburg*)

As mentioned above, the appearance of the Polish knights in this battle was characterized by the bascinet helmet, with or without a visor, arm and leg protection made of plates, the *Lentner* (lamellar armor) and a short tunic over it, and often no shield and no caparisons. Among the Poles, it appears that red clothing with white decorations was preferred.

In contrast to the Poles, the armor and weapons in the Lithuanian Army clearly showed East European influences because each people oriented on their traditions. That was definitely the case with the Ruthenian contingent, in which the Ruthenian (northwestern Slavic) and Tartar warriors wore Eastern or oriental clothing and armor and carried oriental weapons. In the contingents from Kiev, Smolensk and other Slavic principalities, one saw eastern, Mongolian and Tartar-influenced weaponry appearing in the sabers, round shields and eastern body armor that was in general use, and the clothing of well-to-do warriors was rather colorful in its appearance.

On the map of the Polish-Lithuanian Kingdom in 1400, shown in Annex 4, Günter W. Fricke marked the homes of the banners (divisions) with numbers, so that their locations are easy to find.

The Sources on the Polish-Lithuanian Army

The seminal work on the banners of the Polish-Lithuanian and allied armies at the 1410 Battle of Grunwald (Tannenberg) was written by Jan Długosz (1415-1480) in his famous work titled *Annales seu cronicai incliti Regni Poloniae* (Annals or Chronicle of the Illustrious Polish Kingdom). Unlike his work about the Teutonic Order army's banners, the *Banderia Prutenorum*, the *Annales* provided only brief descriptions of the appearance of the 51 Polish and their 40 Lithuanian and allied banners and the devices on them. So over the years, researchers have relied on a number of other sources to expand and refine the information about the standards, using later works about Polish and Lithuanian flags and heraldry (e.g., Długosz's *Insignia sue clenodia Regni Poloniae*, referred to here as "*Insignia*", armorials like the *Codex Bergshammar* (see Annex 10) and the *Armorial of the Knights of the Golden Fleece* (see Annex 9), and depictions on seals, coins, tombs, etc. Added to the complexity of the effort to recreate the banners are the significant differences between Polish and Western European heraldry.

The "Great" Kraków Banner
Figure & photo: W. Friedrich; painted by J. Hensel

"The Battle of Grunwald" by Jan Matejko, 1878, oil on canvas, 426 x 987 cm (14' x 32'), from the collection of the National Museum in Warsaw (MNW), photo: Wikimedia Commons

Polish Heraldry

Unfortunately, there are no drawings of Polish coats of arms in Jan Długosz's *Insignia seu clenodia Regni Poloniae*, but he very accurately describes the coats of arms of Polish clans and districts. Depictions of many Polish coats of arms were, however, in the *Codex Bergshammar* from the second half of the 15th century, the *Armorial of the Golden Fleece* from the late 14th century as well as others from that era. The Polish heraldic system differed from that of western European countries where individual noblemen or their families had their own coat of arms. In Poland, an entire clan would use the same coat of arms, where the clans were originally formed by a group of warriors not necessarily connected by descent. Thus in Poland, there are only 200 known coats of arms as opposed to a country like Germany which has approximately 200,000. At that, only a few dozen coats of arms are documented for medieval Poland. For example, about 450 families of various names were under the Lubicz coat of arms, while there were 200 under Łabędź and 250 under Korczak (images below). These "proclamation" coats of arms had their own names, which were mostly derived from assembly calls or battle cries. But they could also be related to the figure on the coat of arms itself, e.g., *Starykoń*, i.e., "old horse" (see pages 58, 66-67 and 103). Some prominent Lithuanian and Ruthenian nobles were also using Polish coats of arms on their seals by 1410, as evidenced by the similarities of their banners with those of the Poles.

King of Poland from the *Armorial of Knights of the Golden Fleece* (source: herbyzbliska.com)

Polish Coats of Arms

The heraldic devices used in Poland were highly varied and some can be compared with those in western Europe of the time. A number of the devices, such as stylized arrows, crescents, horseshoes, etc., seem to have been derived from Sarmatian property marks, called *tamgas*. For Lithuania, there were two common devices used on banners, a knight and the "Pillars" or "Columns" of the "Gediminids" (image below). The mounted knight in white, holding a sword and shield, is known as "Vytis", also called the Pogoń standard (see pages 52, 55 and 72). The Pillars or Columns of the Gediminids were undoubtedly used on the coat of arms of Grand Duke Vytautas of Lithuania as early as 1397 and were featured on coins of the 14th and subsequent centuries. The symbol, like many of the Polish devices, is believed to have been derived from property markings. See Annexes 8, 9 and 10 for examples.

Pillars of the Gediminids from the 16th century. Copy from Jan Długosz's *Insignia seu clenodia regis et regni Poloniae*

As mentioned above, various military historians calculated that about 20,000 men served in the 51 Banners (divisions or units under a banner) of the Polish Crown and about 10,000 under the 40 Banners of the Lithuanian-Ruthenian forces. This book will not be able to address all 91 banners (divisions), but it should be noted that apparently many banners carried the same flags, e.g., the Pogoń, the Pillars of the Gediminids standards or the "Sun" banners of the Podolian District.

The most prominent modern source on the banners of the combined Polish-Lithuanian army and its Ruthenian allies is probably *Banderia Apud Grunwald I: Chorągwie polskie pod Grunwaldem – Polish Banners at Grunwald* co-authored by Andrzej Klein, Nicholas Sekunda and Konrad A. Czernielewski and published in Łódź, Poland in 2000.

Auksztota / Lithuanian Banner with the Pillars of the Gediminids – Golberg's Go2892
Painted & photo by B. Grimm

Nadworna (Court Band) Banner with V*ytis* - W. Friedrich's engraved flat
Painted & photo by P. Blawatt

That work relied on extensive research of armorials, seals and other sources to reconstruct the appearance of the banners. Two noteworthy earlier works done in the flat tin figure collecting community were by Günter W. Fricke and Heinz Schlenze. In 1972, Günter W. Fricke wrote two articles that were combined and in 1985 published as "*Die Banner des polnisch-litauischen Heeres in der Schlacht bei Grunwald 1410*" ("The Banners of the Polish-Lithuanian Army at the 1410 Battle of Grunwald") in the East German publication *Arbeitsmaterial Kulturgeschichtliche Zinnfiguren* (Cultural-Historical Tin Figures Working Materials) of the Cultural Union of the GDR *(Kulturbund der DDR)* (1985) and written with input from Andrzej Klein who is mentioned above. Heinz Schenzle's 1982 book, *Schlacht bei Tannenberg 1410 Kampf um die Marienburg* [Battle of Tannenberg 1410 Fighting for/around the Marienburg] was published in concert with the German KLIO *Zinnfiguren* collectors' group. The depictions of the banners in these three works were used in the painting of figures for this book, with priority given to Klein's book as it benefitted from many more years of research. However, the figures produced by Hans Stoll

Banderia Apud Grunwald I by Klein, Sekunda & Czernielewski

predated Klein's work and were engraved using Heinz Schenzle's and Günter Fricke's designs. The flags and coats of arms can be seen in the annexes.

W. Friedrich's flat, engraved	The "Great" Kraków Banner	W. Friedrich's flat, engraved
Painted & photo: R. Sanders;	H. J. Stoll's RTP 11, engraved	Painted & photo: Stefan Wachter
	Painted & photo: R. Sanders	

The Flat Tin Figures

With regard to *Zinnfiguren* (flats), the primary producers of Polish and Lithuanian standard-bearers have been Hans Jörg Stoll of Überlingen, the former Golberg International firm (now Wilfried Dangelmaier), Frank and Ivonne Dittmar's Schmalkalder Zinnfiguren, as well as Wolfgang Friedrich of Rackwitz, Saxony. There are additional figures that were produced by Hans Müller of Erfurt in the 1930s. Also Werner Kästner edited 40 figures with engraved Teutonic Order banners for Tannenberg (now available from Schmalkaldener Zinnfiguren), whose emblems can be removed in order to use the flags as Polish-Lithuanian standards. There are also three additional Polish and one Lithuanian mounted standard-bearer figures (TBM 23-25 and TBM 47) available from Schmalkalder Zinnfiguren. If the banner of the figure shown here was engraved with the device, then the caption is marked with "engraved". Unless noted otherwise on the caption or text about a specific figure or banner, the designs will have been based on the Klein book, although all of the engraved Stoll figures are based on Fricke's and Schenzle's designs.

The Banners of the Royal Polish Army

This part of the book will present the banners generally in the order that they were addressed in Długosz's *Annales*, the sequence also followed in Klein's book.

According to historian Wolfgang Büche, "*Długosz is only known to have provided measurements in 'ells' for the Teutonic Order's standards, so one can make one's own decisions about the sizes with the Polish banners. Based on heraldic rules of later times we know that the size of standards was dependent on the importance of the leader of the division. So the main Polish banner (Kraków) was definitely the largest. Whether the other banners from the individual parts of the country were bigger than those of the Polish barons is quite questionable. Perhaps those of the Dukes of Mazovia were larger than those of the barons. These are all questions that cannot be answered. Here only analogies can be drawn from the Order's banners or those of the captured Burgundian flags in Switzerland.*"[106]

The first banner described by Długosz was the **"Great" Kraków banner**, as the principal standard of the Polish-Lithuanian army. During the battle, the banner momentarily fell during an attack on it by a group of Teutonic Knights under the command of the Grand Master himself (who died in the event). This flag's falling is reputed to have caused a temporary crisis in the combined army, but some historians question whether this event ever took place. The scene has been depicted by Wolfgang Friedrich with a figure of the Polish standard-bearer Marcisz z Wrocimowie on his fallen horse, the Grand Master reaching for the banner (see page 50). In the battle, Zyndram z Maszkowic was the head of the Great Camp of the Crown and the Commander of the Banner of Kraków, that was composed of the elite units and carried the banner of the entire army.

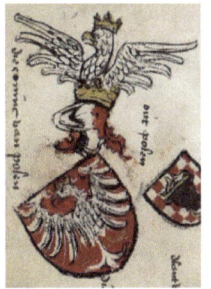

Royal coat of arms from the *Codex Bergshammar*

The "Great" Kraków Banner
Golberg's Go2858, painted & photo: R. Sanders

This banner was a gonfanon, meaning its device was positioned sideways rather than upright. The royal crest can be seen in the *Codex Bergshammar* (see image above at right). While Mr. Friedrich's Marcisz z Wrocimowie figure (page 53, upper left) depicts it as a gonfanon, another of his figures as well as those by Hans Jörg Stoll and Golberg show the Polish eagle upright with a crown (page 53 center and right). The figure, however, displays the coat of arms of the Polutosza clan on its caparison.

While there is no definitive information about the **Polish royal pennon**, it has been suggested that it was also in the form of a gonfanon and bore the crowned Polish eagle in white. Hans Müller of Erfurt edited a figure drawn and engraved by Ludwig Frank sometime around 1937 of a Polish knight with a gonfanon pennon with the eagle engraved. Wolfgang Friedrich's dismounted Teutonic knight with a pennon is also very suitable for painting as the royal pennon (after Andrzej Klein). This small royal banner could have been the standard of the King's bodyguard that consisted of 60 lances.

Royal pennon
H. Müller Erfurt's MP23, engraved
Painted & photo: G. Bistulfi

W. Friedrich's figure
Painted & photo: B. Grimm

The **Banner "Gończa"**, which Długosz listed second, carried a flag described as *"two yellow crosses on a blue field"*, which appears to have been Polish King Jagiello's personal badge. This crest can be found in the *Codex Bergshammar*. This banner was earmarked for the opening of the battle, for pursuits and the vanguard. German sources refer to it as the *"fliegende Rotte"* or vanguard banner. The author has not identified any editors who produced a figure with the Gończa engraved on a standard.

'Gończa' coat of
arms from the *Codex
Bergshammar*

'Gończa' Banner
Golberg's Go2891
Painted & photo by R. Sanders

The third banner in Długosz's *Annales* was the **"Nadworna" ("Court") Banner** that bore the knight on a white horse on a red field. The reconstruction of the *Pogoń* rider in Andrzej Klein's book was based on the sculpture on Jagiello's sarcophagus. The device is also shown in the *Codex Bergshammar* (see below).

At Grunwald, the Banner (division) under this standard was composed of members of the royal court and it was positioned toward the rear of Jagiello's order of battle. There are flat figures by Wolfgang Friedrich, Golberg International, Hans Jörg Stoll and Krista & Wolfgang Wohlmann which are engraved with the Pogoń, all showing the standard-bearer wearing the Lithuanian style war hat.

Nadworna ('Court' or 'Court Band's') Banner

from the *Codex
Bergshammar*

W. Friedrich's, engraved
Painted & photo by H. Meißner

K. & W. Wohlmann's Rus 25,
engraved
Painted & photo by R. Sanders

H. J. Stoll's design
(defunct RTP 2c)

According to Jan Długosz's *Annales*, "*The fourth* **Banner of Saint George** *had as its badge a white cross on a red field*" (note: a reversal). There is considerable controversy over this standard, e.g., whether the colors should be reversed to a white cross on a red field, about the width of the cross, and whether the flag had a tail. In any case, at the battle, it flew over Bohemian and Moravian mercenaries under Jan Sokol. The *Codex Bergshammar* also shows a red shield with a white cross (see Annex 10, the page marked "112" at the lower right at page 108). A figure of this banner was produced by H. Stoll, engraved with a narrow cross. Bernt Grimm has painted a Stoll figure with a pennon in the alternate color scheme and he displayed the *Strzegomia* clan symbol on the horse cover.

St. George Banner
H. J. Stoll's RTP 21c, engraved   Golberg's Go2893
Painted & photos: R. Sanders

St. Georg pennon (alternative colors) Stoll's RTP 8b
Painted & photo: B. Grimm

"*The fifth Banner was that of the* **Poznań District** *having as its badge a white eagle on a red field without a crown*", according to Długosz's *Annales*. The absence of the crown is a key feature distinguishing this standard from the "Great" Kraków banner with a crown, which was the principal standard of the Polish-Lithuanian army. This banner (division) was probably stationed on the Crown army's left wing at the battle.

Długosz wrote "*The sixth [banner], of the* **Sandomierz District**, *had a device, on one side of which were three yellow stripes or bars on a red field, while on the*

Poznań District
Stoll's RTP 3c, engraved
Painted & photo by R. Sanders

Sandomierz District
Stoll's RTP 6c, engraved
Painted & photo by R. Sanders

second were seven stars on a sky-blue field." There are contemporary representations of this device and ones from later centuries that show the stars as yellow or white and in varying arrangements, and the bars placed differently. For example, the *Codex Bergshammar* features the stripes in a different sequence and the stars staggered.

Sandomierz   Kalicz
Coats of arms from the *Codex Bergshammar*

The Sandomierz division was stationed on the extreme right wing of the Crown army and participated in the counterattack on the Teutonic Order Army when it was returning from its successful attack on the Lithuanian wing.

Kalisz District
H. J. Stoll's RTP 10c, engraved   W. Friedrich's TAM24   Sieradz District
Painted & photos by R. Sanders   H. J. Stoll's RTP 17b
Painted & photo: B. Grimm

The seventh banner, that of the **Kalisz District**, was described in Długosz's *Annales* as "*having as its badge the head of an aurochs on a checker decorated with a royal crown, from whose nose hung a ring.*" See Annexes 5, 7 and 10 (page marked "112" first row). The figure with this engraved banner was produced by Hans Jörg Stoll and the shield bears the sign of the Belina clan, whose painting is also based on a depiction in the *Armorial of Knights of the Golden Fleece*. The figure's shield and surcoat in the photo above center displays the *"Topór"* (axe) coat of arms that was used by many *Szlachta* (noble) families.

The eighth banner, from the **Sieradz District**, "*had as its badge on one side half a white eagle on a red field, and on the other side half a flaming lion on a white field.*" Długosz's term "flaming" (Latin *flammei*) is ambiguous, and it could have been simply half of a lion's head breathing flames or of a lion rampant as depicted here after Andrzej Klein's depiction. This coat of arms does not appear in either the *Bergshammar* or the *Golden Fleece* armorials.

The **Lubin Banner**, the ninth in the *Annales*, was described as having "*a stag with splayed antlers on a red field*" but in Długosz's *Insignia*, he adds that the stag was standing up straight and was adorned with a golden crown around its neck. While no contemporary depictions of the banner itself survive, a crest with a stag with a crown can be seen in the *Codex Bergshammar*.

Lubin coat of arms from the *Codex Bergshammar*

The location of this division during the Grunwald battle is unknown. Fricke's, Schenzle's and as a result also Stoll's depictions of the flag show a white stag standing on green grass on a red cloth. (See Annex 7, plate 2, No. 16), while Klein presents a variation (Annex 5, No. 9). The Lubin standard bearer's shield (above) displays the Lubicz Clan's symbol.

Lubin District
H. J. Stoll's RTP 12c, engraved
Painted & photo: R. Sanders

Łęczyca District
H. J. Stoll's RTP 18b
Painted & photo by Bernt Grimm

The Banner of the **Łęczyca District**, the 10th in the *Annales*, "*had as its badge half a black eagle and half a white lion on a yellow field with crowns on their heads.*" The depiction of the banner in the figure shown here is from Andrzej Klein's work which was based on the Kuyavian coat of arms. It does not appear in the *Bergshammar* or *Golden Fleece* armorials. Hans Jörg Stoll produced a figure with an engraved Łęczyca banner (RTP 9), shown above with the sign of the *Starykoń* (Polish for "old horse") clan painted on the shield.

Kuyavia coat of arms from the *Codex Bergshammar*

The **Kuyavian District**'s Banner, the 11th in the *Annales*, "*had as its badge half a black eagle on a yellow field and half a white lion on a red field with crowns on their heads.*" While the *Golden Fleece* armorial

Kuyavian District
W. Friedrich's figure
Painted & photo by Horst Meißner

H. J. Stoll's RTP 9c, engraved
Painted & photo: R. Sanders

shows the half black eagle on a white field, the figurines shown below used Długosz's description which also matches the *Codex Bergshammar* image. Hans Jörg Stoll's Łęczyca figure has been painted here as the Kuyavian banner. The figure painted by Horst Meißner (opposite, left) shows the Knight of Sieradz with the "Roja" (rose) device of the Poraj Clan on the shield.

Lwów coat of arms from the *Codex Bergshammar*

Lwów District
Golberg's Go2893
Painted & photo by B. Grimm

H. J. Stoll's RTP 22c, engraved
Painted & photo by R. Sanders

The **Lwów District**'s banner appeared twelfth in the *Annales* and displayed "*a yellow lion walking on a rock on a blue field*" which agrees with that in the *Golden Fleece* armorial (page 105) and the *Codex Bergshammar*. Again, Hans Jörg Stoll has produced a figure with Lwów District's badge engraved on it.

Perhaps the simplest banner attributed to the Polish-Lithuanian Army was that of the **Wieluń District**, simply "*a snow-white transverse band on a red field in equal proportions.*" In a description of the battle,

Stoll's RTP 12b
Painted & photo: B. Grimm

Wieluń District
Stoll's RTP 17c, engraved
Painted & photo: R. Sanders

Stoll's RTP 25b, after G. Fricke's rendering - Painted & photo: R. Sanders this standard was located together with those of the Sandomierz and Halicz Districts and the "Great Banner of the Kraków District" among others and it took part in the counterattack after the Lithuanian wing broke. Two other versions of the Wieluń flag, one bearing a "lamb of God" and the other with two linked horns were shown in Fricke's and Schenzle's works (see Annex 7, plate 1, No. 13) and produced by Stoll; however, the "lamb" banner probably did not originate until the mid-15th century and the origin of the version with the horns is unclear. None of these three devices appear in either in the *Golden Fleece* armorial or the *Codex Bergshammar*. The figure painted by Bernt Grimm shows the "Roja" sign of the Poraj Clan on the shield (see above and Annexes 8, 9 and 10).

The fourteenth banner in Długosz's *Annales* was that from the **Przemyśl District**, described as "*a golden eagle with two heads each turned away one from the other on a blue field*"; however in his *Insignia*, he wrote it was on a red field. No contemporary drawing of the eagle exists, so its appearance was extrapolated. Długosz did not describe this division's role in the battle. The eagle engraved on the banner of Hans Jörg Stoll's figure was based on the Heinz Fricke drawing (Annex 7, plate 2, No. 20); the painting of the device by Bernt Grimm is based on Andrzej Klein's illustration.

Przemyśl District
H. J. Stoll's RTP 19c, engraved
Painted & photo: R. Sanders

Golberg's Go2891
Painted & photo: B. Grimm

In the *Annales* the 15th Banner, that of the **Dobrzyń District**, had an emblem of the face of a "*wild looking man... the head was adorned with a crown and also with horns in a yellow field.*" In Długosz's *Insignia* there is a similar description but saying in a red field, which was the common color in later sources. Fricke's reconstruction is based on the coat of arms in the *Armorial of Knights of the Golden Fleece*. The *Codex Bergshammar* shows two different crests with a king's head, one on a black and one on a blue field but neither may have been intended to represent Dobrzyń (see Annex 10, pages marked "112", bottom row and "290" bottom row). The standard-bearer figure's shield shows the "lily" device of the Gozdawa Clan.

Dobrzyń District
Stoll's RTP 18c, engraved – Painted & photo: R. Sanders; an alternate version by G. Fricke; a coat of arms from the *Codex Bergshammar*

Chelm District
Golberg's Go2890

Podolien District
H. J. Stoll's RTP 27c, engraved

Painted & photos by R. Sanders

The 16th Banner came from the **Chelm (German: Kulm, Polish: Chełmno) District** and its standard *"had a white bear standing between two trees, in a red field"* as described in the *Annales*, while in Długosz's *Insignia* he described it with two trees behind and one in front of the bear as shown here. Other written and graphic sources give additional variations of both the colors and the numbers of the trees.

The 17th, 18th and 19th Banners (divisions) all came from the **Podolian (Podolskiej) District** due to the large number of knights who owned land there, and the three banners fought under the same standards. The banners *"had the sun's face in a red field."* The image on Stoll's flag, based on the coat of arms of Podolskiej, does not correspond to that which is in Klein's book. The field is more often shown as white than as red but a yellow emblem on a white field is rare (see Erwin Ortmann's depiction in Annex 6).

"The Battle of Grunwald" painted by Stefana Garwatowskiego[107]

At the battle, the **Halicz District**'s banner carried a standard that bore "*a black jackdaw* with a crown on its head in a white field....*" The banner's design in the photo to the right is based on Klein's illustration (see Annex 5, p. 93, no. 20) derived from the *Codex Manesse*, while Schenzle and Fricke used a different version shown in Annex 7, sheet 2 (p. 93, no. 21) and which was used by Stoll in one of his engraved figures which is no longer being produced. Another version of the bird can be seen in the *Codex Bergshammar* (above and see Annex 10, p. 108, upper right). It is known that the 20th Banner played a decisive role in the battle of Tannenberg.

A coat of arms from the *Codex Bersama*

Halicz District
H. J. Stoll's RTP 19b
Painted & photo: R. Sanders

Mazovian District
H. J. Stoll's RTP 1b
Painted & photo by B. Grimm

The 21st and 22nd Banners, which are discussed in the *Annales*, "*were those of the Duke Siemowit of Mazovia with a white eagle without a crown in a white field as its symbol.*" The emblem of the white eagle, similar to that of King Jagiello, is derived from this noble's membership of the District of the Piasten. The style in the figure shown here is based on its depiction in the *Golden Fleece Armorial*. These two divisions were actually led by Duke Siemowit's son because the father was supporting the Teutonic Order in the conflict. Bernt Grimm's figure shows the *Korzkiew* device on the shield.

W. Friedrich, engraved
Painted & photo: S. Wachter

Janusz of Mazovia
Figure & photo: W. Friedrich,
Painted by J. Hensel

H. J. Stoll's design of the former
RTP 13c figure

*A European blackbird related to a crow or a grackle.

With respect to the 23rd Banner, which Długosz identified as belonging to the **Duke Janusz of Mazovia**, the standard itself probably belonged to the Czersk District of Mazovia and would have had an eagle and a dragon in white or red on complementary red and white quartered fields. The depiction on the facing page at the left shows the caparison in the ostensible coat of arms of the Duke of Bremoss from Silesia, while the photo in the center, opposite, shows the standard-bearer's caparison with the arrow, the stars and the crescent moon of the Sas Clan in a blue field. Hans Jörg Stoll produced a standard-bearer figure of the Janusz von Mazovia banner (RTP 13c), which is, however, no longer available.

Mikołaj Kurowski
Stoll's RTP 20b
Painted & photo: R. Sanders

Wojciech Jastrzębiec
Stoll's RTP 12b
Painted & photo: B. Grimm

Krystyn of Ostrów, the Castellan
of Kraków - Stoll's RTP 2b
Painted & photo: B. Grimm

"The 24th [Banner] was that of the Archduke **Mikołaj Kurowski of Gniezno**, which had a river with a cross in a red field as its device." The coat of arms, named Szreniawa, is a white reversed "lazy" letter "S", named "krzywasn" with a small yellow cross in a red field. The name "krzywasn" is an Old Polish word for "river", so this was "the Clan of the River". Mikołaj Kurowski was the chancellor of the Polish Kingdom starting in 1394. At Tannenberg he raised his own banner (division). After the battle, he was a negotiator at the Peace of Thorn and negotiations with Sigismund of Luxemburg.

"*The 25th [Banner] was that of **Wojciech Jastrzębiec**, Bishop of Poznań, which had a horseshoe with a cross in the middle in a blue field as its device.*" Wojciech did not personally take part in the battle but his forces were commanded by Jarand of Brudzewo. Hans Jörg Stoll presented a standard-bearer of the Bishop of Poznań (Posen) but the figure (RTP 14c) is no longer available.

"*The 26th [Banner] was that of **Krystyn of Ostrów**, the Castellan of Kraków, with its device being a bear carrying a wreathed virgin in a red field.*" This coat of arms represented the Rawa, Rawicz or Rawiez Clan, whose members were most frequently found in the Kraków and Sandomierz Districts.

Szreniawa Jastrzębiec Rawicz
Coats of arms from the undated *Codex Bergshammar*

The **27th, 37th and 39th Banners** had identical standards with a yellow star over a yellow crescent in a blue field. The divisions were those of **Jan of Tarnów** (Voivode of Kraków), **Wincentego of Granowo** (Castellan of Naklo and Starosta-General of Greater Poland - Wielkopolska) and **Spytko of Jarosław and Tarnów** (the Starost-General of the Ruthenians and Voivode of Sandomierz). A *Starosta* was the main administrator of a Polish district or region. The device is that of the Lilawa Clan with its seat in east-central Poland. A figure with an engraved Lilawa Clan coat of arms (RTP 24c) was produced by Hans Jörg Stoll (above left).

Lilawa Clan – flag of 27th, 37th and 39th Banners
H. J. Stoll's RTP 24c, engraved
Painted & photo: R. Sanders

Sędziwój of Ostroróg and Dobrogost Świdwa of Szamotuli
H. J. Stoll's RTP 23b
Painted & photo: B. Grimm

The standard shown above on the right would have flown over the **28th and 41st Banners** and borne a neckerchief rolled or twisted into a circle in a red field. The neckerchief could have been white or yellow. The 28th Banner was under Sędziwój of Ostroróg, the Voivode of Poznań and the 41st Banner was led by Dobrogost Świdwa of Szamotuli, the later Chamberlain of Kalisz. The two commanders belonged to the Nałęcz Clan whose symbol appeared on their standards.

Left: Mikołaj of Michałów
H. J. Stoll's RTP 24b
Painted & photo: R. Sanders

Right: Jakub of Koniecpol
H. J. Stoll's RTP 5b
Painted & photo: B. Grimm

The 29th Banner was that of **Mikołaj of Michałów**, the Voivode of Sandomierz, and its "Roja" coat of arms was a white rose in a red field, here after its depiction in the *Armorial of Knights of the Golden Fleece*. The *Codex Bergshammar* shows a crest that can give a different style for such a rose (see the facing page).

The **Banner of Jakub of Koniecpol**, the Voivode of Sieradz, was the 30th Banner. He was also the *Starosta* of Kuyavia at the time of the battle and his device was the *Pobóg*, a horseshoe with a cross on it (see facing page). The *Pobóg* was used by many noble families in Poland, sometimes on a blue field and sometimes with the cross in yellow or gold, sometimes also with the opening shown to the top.

Iwo of Obichów,	Jan Łęczyca of Bobrek	Andrzej of Tęczyn
the Castellan of Szremsk	H. J. Stoll's RTP 23b	H. J. Stoll's RTP 11b
Golberg's Go2891	Painted & photo: R. Sanders	Painted & photo by R. Sanders
Painted & photo by B. Grimm		

In the *Annales* it is written that: "*The 31st [Banner] was that of Jan or **Iwo of Obichów**, the Castellan of Śrem, who had as his device a bison's head with a golden ring passed through its nose in a yellow field.*" It is the coat of arms device of the Wieniawa group of families. The depiction of the bison's head in the image on the left is based on similar depictions, shown in the *Armorial of Knights of the Golden Fleece*.

At the Battle of Grunwald, the 32nd Banner was personally commanded by **Jan Ligęza of Bobrek**, the Voivode and *Starosta* of Łęczyca. His standard bore the Półkozic coat of arms, which displayed the head of an ass in black in a red field. The *Codex Bergshammar* shows what may be this device.

"*The 33rd [Banner] was that of **Andrzej of Tęczyn**, the Castellan of Wojnice, that had an axe in a red field as its device.*" This device, named the *Topór*, was used by the Starża Clan, to which he belonged. The *Topór* coat of arms (Polish for "axe") was used by many *Szlachta* (noble) families which are sometimes called "*Starża*", an Old Polish word that indicates advanced age. The *Topór* coat of arms would have been found on the shields of many Poles at Grunwald (as on the flag in the image above on the right).

  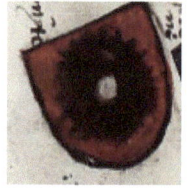 

| Lilawa Clan | Nałęcz | Roja | Pobóg |

 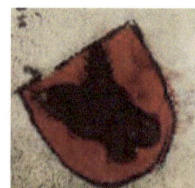 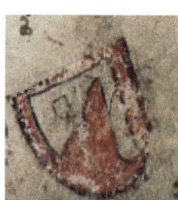

| Pomian | Półkozic (?) | Topór |

Coats of arms from the undated *Codex Bergshammar*

In the *Annales* it says: *"The 34th [Banner] was that of Zbigniew z Brzezia, **Marshal of the Polish Kingdom**, whose device was a lion's head spewing a flame, in a blue field"* (image to left). This coat of arms belonged to the Zadora Clan which possessed lands around Kraków and Sandomierz and along the Pilica River. The Marshal of the Polish Kingdom was the most important military leader after the King himself and he rode at the van of the army where the royal pennon was normally carried (see page 54). During the Battle of Grunwald, Zbigniew z Brzezia personally took over command of a group of four to six banners, including the Gończa and the St. George Banners as well as his own, which formed the screen between the Lithuanian forces and the right wing of the Royal Army.

Marshal of the Polish Kingdom
Stoll's RTP 5c, engraved
Painted & photo: R. Sanders

Piotr Szafraniec of Pieskowa Skała
H. J. Stoll's RTP 28b
Painted & photo: R. Sanders

The image above on the right shows the standard-bearer of the 35th Banner under **Pjotr Szafraniec of Pieskowa Skała**, the governor of Podolia starting in 1404, the Chamberlain of Kraków and after 1408 also the Marshal of the Court. The flag's emblem was that of the *Starykoń* (Polish for "Old Horse") Clan, *"a white horse, with a black girdle in its middle, in a red field"*, a coat of arms that was used by many *Szlachta* (noble) families among the Polish-Lithuanian national union.

*"The 36th [Banner] was that of **Klemens of Moskorzew**, the Castellan of Wiślica, whose symbol was two and a half yellow crosses in a blue field"* according to the *Annales*. Klemens was earlier the Vice Chancellor of the Polish Kingdom and *Starosta* of Kraków (photo at right).

The **37th and 39th Banners** were described above with the 27th Banner of the Lilawa Clan on page 64.

Klemens of Moskorzew
K. & W. Wohlmann Got7
Painted & photos by R. Sanders

Dobiesław of Oleśnica
using Wolfgang Friedrich's "Bischof von Menz" figure
Painted & photo: B. Grimm

The photo on the facing page, lower right, shows the standard of the 38th Banner, that led by **Dobiesław of Oleśnica**, as it was described in the *Annales*. According to some reports, even though he was already 50 years old, he personally attacked Grand Master Ulrich von Jungingen at Grunwald. Dobiesław was the Castellan of Wojnicki, Lublin and Sandomierz; the *Starosta* of Kraków; the *Podczaszy* of Kraków; and the Voivode of Sandomierz. This Dębno coat of arms appears in the *Codex Bergshammar* on the page now marked "290" in the center row and is shown below.

Długosz wrote: "*The 40th [Banner] was that of **Marcin of Sławsko**, which had as its device a lion's body in its upper half and four jewels in its lower half.*" The coloring of this "Zaremba" coat of arms and the arrangement of the jewels varied considerably from that shown here at the left, so that the upper part was yellow and the lower part red, and the four jewels are shown in various colors. Two examples can be seen in the *Codex Bergshammar*, shown below. Another depiction of the Zaremba coat of arms, based on the *Armorial of Knights of the Golden Fleece*, from Andrzej Klein's *Apud*, can be seen in Annex 5 as No. 40a. At the time of the Battle of Tannenberg/Grunwald, this banner's leader, Marcin of Sławsko, was the Deputy Master of the Pantry for Poznań.

The **41st Banner** was discussed above with the 28th Banner.

"*The 42nd [Banner] was that of **Krystyn of Koziegłowy**, the Castellan of Sącz, which had as its device an arrow decorated with a double cross in a red field*", as it was described in the *Annales*. This reconstruction of the coat of arms is based on one of the devices shown in the *Armorial of Knights of the Golden Fleece*.

Marcin of Sławsko
Stoll's RTP 19b
Painted & photo: B. Grimm

Krystyn of Koziegłowy
Werner Scholtz's 1143
Painted & photo: R. Sanders

| Marshal of the Polish Kingdom | *Starykoń* | Dębno | Zaremba (40) | Zaremba (40a) | Koziegłowy or Sącz |

Coats of arms from the undated *Codex Bergshammar*

Jan Mężyk of Dąbrowa
W. Friedrich
Painted & photo: H. Meißner

Mikołaj Trąba of Wiślicz
H. J. Stoll's RTP 20b
Painted & photo: R. Sanders

In the *Annales* by Długosz it reads: "*The 43rd [Banner] was that of **Jan Mężyk of Dąbrowa** with two fishes, which were called trout, one in a white field, the second in a red field.*" This reconstruction is based on a flag in the *Banderia Prutenorum*. The flat tin figure shown here, painted with the fishes emblem, was created by Wolfgang Friedrich. It was painted by Horst Meißner as the banner of the "Knight von Wadwicz" with the caparison's Łabędź (Polish for "swan") coat of arms that was widespread in the Kraków and Sandomierz Districts.

The image above right shows the standard-bearer of **Mikołaj Trąba of Wiślicz**'s (the 44th) Banner, which is described in the *Annales*. The flag had "*three horns in a white field*" and was used by the "*Trąba*" ("Horn") Clan. This Division was raised by Mikołaj Trąba of Wiślicz who at the time of the battle was the Vice Chancellor of the Polish Kingdom and the Archbishop of Halicz. This rendering's style is very similar to that of a coat of arms in the *Codex Bergshammar*, shown below, but on a red field.

"*The 45th [Banner] was that of **Mikołaj Kmita of Wiśnicz** with a red river decorated with a cross as its device.*" Like the (24th) Banner of Mikołaj Kurowski, Archduke of Gniezno, this is the *Szreniawa* coat of arms but in reversed colors. It is probable that Długosz made a mistake with these colors.

According to Długosz, "*the 46th [Banner] was that of the **Knight Brothers of the Griffin (Gryf)** that had a white griffin in a red field as its device.*" This was one of the two banners under which the clan fought at Tannenberg. This reconstruction in the photo opposite is based on a coat of arms in the *Armorial of Knights of the Golden Fleece*. It also appears in the *Codex Bergshammar* as shown here.

Trąba   Gryf
Coats of arms from the *Codex Bergshammar*

Mikołaj Kmita of Wiśnicz
H. J. Stoll's RTP 27b
Painted & photo: R. Sanders

Knight Brothers of the Griffin
(*Gryf*) - Stoll's RTP 4c, engraved
Painted & photo: R. Sanders

Hans Jörg Stoll used the Heinz Schenzle/Günter Fricke design for his figure's engraved flag (facing page).

The 47th Banner was that of the **Zaklika of Korzkiew** knights who had a coat of arms with a white "W" with a cross above it in a red field. This symbol can be found on the *Codex Bergshammar* sheet marked "291" center row, fourth crest. According to Polish chronicles about the war against the Teutonic Order, Zaklika of Korzkiew was the knight who informed King Jagiello about the Order's defeat at the October 10, 1410 Battle of Koronowo (German name: *Polnisch Krone*).

Zaklika von Korzkiew
Golberg's Go2894
Painted & photo: R. Sanders

Jelita Clan
Werner Scholtz's 1054
Painted & photo: R. Sanders

The Jelita Clan's standard shown in the image above right is that of the 48th Banner, *"the Knight Brothers of the Koźle Rogi Clan with three crossed lances in a red field,"* so the *Annales* tells us. The **Jelita Clan**'s division was the second banner assembled in the summer of 1410 during preparations for the Grunwald campaign, in which a single clan fought together in one banner. The Koźle Rogi ("Goat Horns") clan used a device with three lances in a red field that was called *Koźlarogi* or *Jelita*. The device from the *Codex Bergshammar* can be seen here. The wealthy clan owned lands around Piotrków, Radom and Kielce.

*"The 49th Banner was that of the **Moravian Baron Jan of Jičín** (Jana z Jičína), that had as its device a white arrow with a curve on its end, which the Poles called 'Odrowąź'"* (see image from the *Codex Bergshammar*.). Długosz wrote the banner was commanded by a Moravian and that it was a mercenary unit in which only Moravians served. During the Battle of Grunwald, this banner together with the other Czech mercenaries (50th Banner) were stationed behind the Banner of St. George and helped the latter re-assemble after their infamous retreat.

Banner of Jan of Jičín (Moravian mercenaries) - Stoll's RTP 22b -
Painted & photo: R. Sanders

Zaklika   Jelita   Odrowąź

Coats of arms from the *Codex Bergshammar*

The *Annales* report: *"The 50th [Banner] was that of **Gniewosz of Dalewice**, the deputy to the Chamberlain of Kraków, who had as his device an arrowhead that had a bar from left to right in the middle, over a two-armed fork standing in a red field."* This device can be found in the *Codex Bergshammar*, as shown on the facing page. This detachment consisted of mercenaries from Bohemia, Moravia and Silesia, who were hired by Gniewosz and who possibly had clan relationships in these countries. As already mentioned, at Grunwald this detachment helped gather the Banner of St. George. Stoll made a flat tin figure with an engraved flag with the "Kościesza", "Strzegomia" or "Strzegomya" coat of arms (above left).

Moravian Banner of Gniewosz of Dalewice / mercenaries
Stoll's RTP 23c, engraved
Painted & photo: R. Sanders

Silesian & Bohemian mercenaries
Stoll's RTP 13b, after Heinz Schenzle's "B8" flag
Painted & photo: B. Grimm

Günter Fricke shows a different "Strzegomia" symbol for the banner of the **Silesian as well as Bohemian-Moravian mercenaries** in his plate (see Annex 7, plate 1, No. 8), while Heinz Schenzle presents that (horseshoe) symbol on the flag under the command of Gniewosz of Dalewice, whom he identified as the Steward of Kraków. The flag here is painted by Bernt Grimm.

Günter Fricke and Heinz Schenzle both listed and drew a banner (see Annex 5, No. 31) under the Polish-Lithuanian participants as belonging to the **Bishop of Kraków**, that is depicted with a white standard with three crowns in yellow or gold. According to Heinz Schenzle, the three crowns stand for Kraków and the Bishoprics of Vilnius and Żmudź in Lithuania.

Günter Fricke also indicates that a unit from the **Principality of Mazovia**, which carried a standard with a white bull's head and a yellowish gold rose, star and crescent, in a red field (Annex 5, No. 24), was part of the forces from the Podolia District. The Duchy of Maz-

Bishop of Kraków
H. J. Stoll's RTP 11b (after G. Fricke)
Painted & photo: B. Grimm

Banner of the Prince of Mazovia
(after G. Fricke)
H. J. Stoll's RTP 20c, engraved
Painted & photo: R. Sanders

ovia (Polish: Księstwo Mazowieckie) came into being as a feudal state in the course of the 12th century on the territory of present-day Mazovia. After the inheritance partition of the Kingdom of Poland in 1138 through the Will and Testament of Duke Bolesław III "Wrymouth", Mazovia fragmented into partial duchies. The origin of Günter Fricke's information about this detachment is unclear. Hans Jörg Stoll used the Günter Fricke illustration for his figure's engraved flag.

Dalewice coat of arms from the *Codex Bergshammar*

"Banners of Kingdom of Poland and Lwów" by Arthur Orlionov

The Lithuanian Contingent and its Banners

Lithuanian *Pogoń*
W. Friedrich's, engraved
Painted & photo by S. Wachter

Duke Zygmunt Son of Korybut
K. & W. Wohlmann's Rus 25
Painted & photo by B. Grimm

In the *Annales* it states: "*The 51st [Banner] was that of the Lithuanian Duke Zygmunt, Son of Korybut, which had as its device a horse carrying an armed man in a red field.*" We know now that this Pogoń device could appear in fields of varying colors (as a rule red or blue) and the horse's color and its furniture could vary. It is suspected that this banner (unit) was made up of troops that were hired by Duke Zygmunt (Sigismund) that had flags in this pattern in varying color schemes. It is possible that Długosz listed this banner as the 51st in order to associate it with the other banners that were manned with mercenaries.

Auksztota / Lithuanian Banner
Golberg's Go2892

Gates or Pillars of the Gediminids
H. Müller's MP27
Painted & photo: R. Sanders

Długosz wrote in his *Annales* that there were 40 banners under the command of the Lithuanian Grand Prince Vytautas in which there were Lithuanians, Samogitians[108] and Tartars serving. Długosz indicates that they were not so well armed or mounted as the men in the Polish banners. Thirty of their banners (units) had the Pogoń on their standards – "*an armored man on a white, black, red or spotted horse, who swings a sword in a red field.*" The ten remaining banners "*had the symbol with which Vytautas branded his horses*" in a red field, followed by a drawing of the Pillars of the Gediminids. It is rather clear that Długosz believed that all of the banners of the Lithuanian Army had either the Pillars or the Pogoń device. The Pillars are often depicted in silver or gold but always in a red field.

Günter Fricke and Heinz Schenzle presented an additional standard, for **Braclaw** (today Bratslaw in Ukraine), which at the time of Grunwald was part of the Lithuanian lands (Annex 7, plate 6, No. L XI). Hans Jörg Stoll produced a figure with its emblem, a blue field, with a red shield, with a white cross on which is a smaller blue shield with a white crescent, after Günter Fricke's illustration No. L XI in Annex 7, plate 6.

In Günter Fricke's work he presents a banner for **Żmudź**, the Polish name for Sambia, today a region in Lithuania. The banner shows a black bear, wearing a golden collar with a ring, in a red field. Hans Jörg Stoll also produced a Żmudź standard-bearer with an engraved banner but the figure is now only available without the coat of arms on the banner. He used the illustrations by Günter Fricke (Annex 7, plate 5, No. L IV and Heinz Schenzle (Li7) as reproduced on the right above.)

Braclaw
H. J. Stoll's RTP 25c, engraved
Painted & photo: R. Sanders

Żmudź
H. J. Stoll's RTP 26b, nach Schenzle
Painted & photo: R. Sanders

In his article, Günter Fricke presents a red standard with a white orthodox cross (Annex 7, plate 6, No. L IX), which is to represent Lithuanian **Kremienice**, probably a town in the present northwest of Ukraine. He gives no further explanation for including it in his article but it is shown here with a figure by Hans Jörg Stoll.

Among the standards presented by ROFUR is one from **Samogitia** with a walking black bear in a red cloth with three tails. This flag could have belonged to the forces under Grand Duke Vytautas of Lithuania. The glued-on banner is shown here on the figure of a Lithuanian horseman.

Krzemienice
H. J. Stoll's RTP 22b
Painted & photo: R. Sanders

Banner from Samogitia
Schmalkalder Zinnfiguren TBM24 (with a ROFUR flag)
Painted & photo: R. Sanders

The Ruthenian Contingent and its Banners

Despite the assertion by Długosz, it is unlikely that either the Ruthenian (e.g., **Smolensk or Kiev**) or the Tartar forces simply fought under the Pillars of the Gediminids or the Pogoń banners instead of bringing along their own standards. The Smolensk District raised three banners that fought together on the outermost left wing of Vytautas's army. The standard to the right is based on a depiction by Heinz Schenzle, while the two standards shown below were probable versions proposed by Andrzej Klein in his *Banderia Apud Grunwald*.

Smolensk
K. & W. Wohlmann's Rus 25b, after Günter Fricke
Painted & photo: R. Sanders

Smolensk
K. & W. Wohlmann's Mosk1
Painted & photo: R. Sanders

Smolensk
Golberg's Go2892 (flag converted)
Painted & photo: B. Grimm, after Andrzej Klein (p. 113)

Hans Jörg Stoll also produced a Kievan standard-bearer (RTP 31c) that is, however, no longer available. Figures produced by Krista and Wolfgang (†) Wohlmann, like their Rus27, are suitable for representing this banner (see Annex 7, plate 5, No. L VII).

Kiev
Left: K. & W. Wohlmann's Rus27 painted & photo by R. Sanders;
right: H. J. Stoll's RTP 31c "*Kiew*" Banner design

A flag presented by Günter Fricke and Heinz Schenzle is that from **Czernikow** (Annex 7, plate 6, No. L X) as part of the Lithuanian contingent but according to Fricke's map, he shows it in the vicinity of Kiev, Ukraine. So it must be a Ruthenian unit from Chernihiv, northeast of Kiev. He did not tell how this unit relates to the battle if, it even was present. This unit is not mentioned in Andrzej Klein's work.

Czernikow
H Stoll's RTP 25b
Painted & photo: B. Grimm

The Tartars and Their Standards

The Tartars who fought as part of Grand Prince Vytautas's army at Tannenberg/Grunwald against the Teutonic Order were commanded by Dzalal el-Din, an exiled Khan. There are no preserved descriptions of the Tartar banners from the battle but based on medieval Persian and other sources it is possible to extrapolate how their standards appeared, often as natural colored or dyed horsetails and colored cloths, probably in the traditional black, green and white. The shape of the cloth could vary considerably. The figures to the right and below show some likely versions of Tartar standards.

A Tartar banner
Golberg's Go1337.
Painted & photo by
R. Sanders

W. Friedrich, engraved
Painted & photo by S. Wachter

Tartar banners
Ulrich Lehnhart figures
Painted & photos by Peter Blawatt

In Conclusion

A little over 600 years ago, the largest battle of the Middle Ages took place at Tannenberg (in East Prussia) in what now is Poland, or at Grunwald as it is called in Polish. What we know about this historically pivotal battle, which spelled the beginning of the decline of the Teutonic Order's state, comes primarily from documents written by the Polish victors, primarily by Jan Długosz (1415-1480). The badly routed Germans seem to have had little interest in writing about their own embarrassing defeat. So ironically what we know about both armies and the actual appearance of their flags at the battle is very detailed about the Teutonic Order's standards captured by the Polish-Lithuanian army while what we know about how the Polish-Lithuanian flags appeared is very limited. And for the Teutonic Order forces, many of the standards are misidentified but such is often the case with war trophies. Conversely the accuracy about which Polish-Lithuanian "Banners" carried specific flags is more detailed, but there are no known contemporary depictions of their battle standards, just short narratives by Jan Długosz describing them. So for the appearance of the combined Polish-Lithuanian and their allied contingents' flags, reconstruction using many sources and much conjecture has been necessary.

Simply put, we may never know the full story of the Tannenberg flags but for figure collectors and wargamers there are a host of fine figures available. For *Zinnfiguren* collectors, many of the figures come with banners engraved with the crests. There are also several sources of flags that can be printed and attached to both "flat" and 3D figures to depict the momentous events of 1410. Hopefully this book will provide a helpful resource in the absence of many earlier works that are no longer in print.

Polish-Lithuanian Banners produced by Hans Jörg Stoll
Painted & photo: R. Sanders

ANNEXES

Annex 1: Banners of the Teutonic Order and its Allies for Attaching

Dignitaries (*Grossgebietiger*), Bishoprics, and Dukes

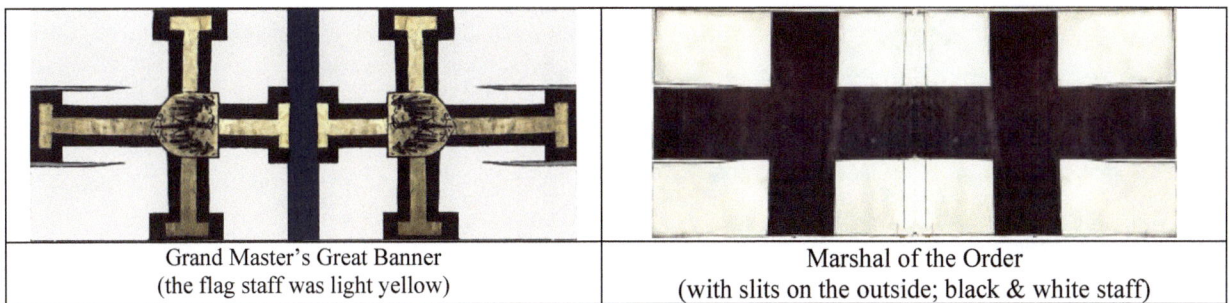

Grand Master's Great Banner (the flag staff was light yellow)	Marshal of the Order (with slits on the outside; black & white staff)

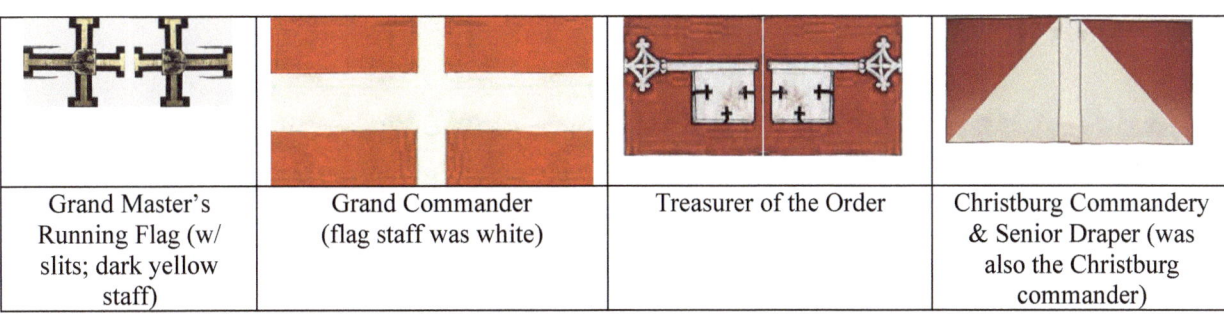

Grand Master's Running Flag (w/ slits; dark yellow staff)	Grand Commander (flag staff was white)	Treasurer of the Order	Christburg Commandery & Senior Draper (was also the Christburg commander)

Bishopric of Pomesania	Bishopric of Sambia (*Samland*)	Bishop of Warmia (*Ermland*)

Duke of Oels (staff w/ equal black & white vertical lines)	Duke of Stettin

Commanderies, Cities and Towns

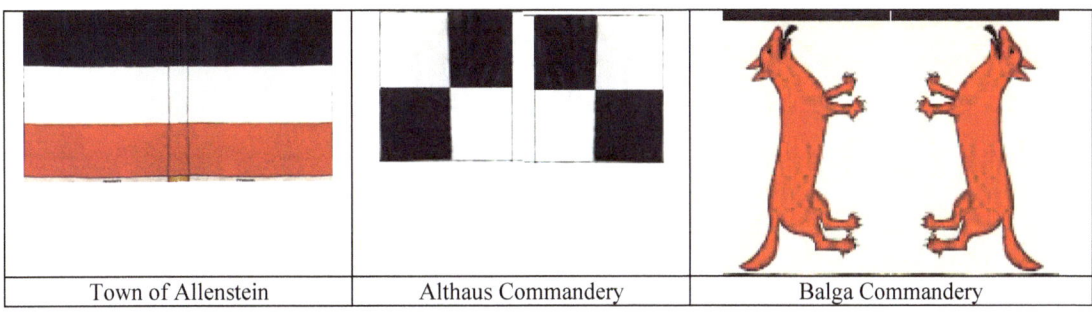

Town of Allenstein	Althaus Commandery	Balga Commandery

 Bartenstein Treasurer's Office (an Order's caretaker sat here)	 Town (and/or Commandery?) of Brandenburg
 Brattian Stewardship	 Town of Braunsburg (light brown staff)  City of Danzig

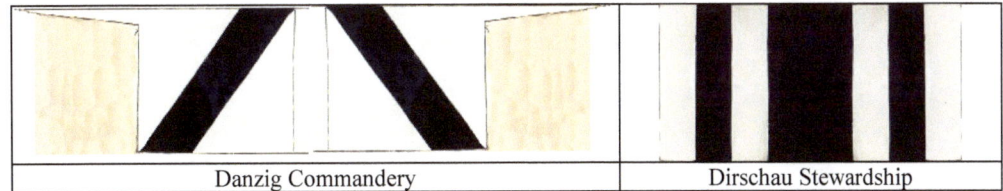

Danzig Commandery	Dirschau Stewardship

 Commandery Elbing's House Commander	 Elbing Commandery and Senior Hospitaller	 Town of Elbing - possibly Długosz switched colors	 Engelsburg Commandery (light brown staff)
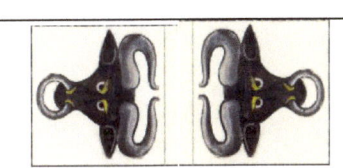 Graudenz Commandery	 Leske Stewardship		 Town of Heiligenbeil
 Old Town (*Altstadt*) Königsberg		 Königsberg Commandery	

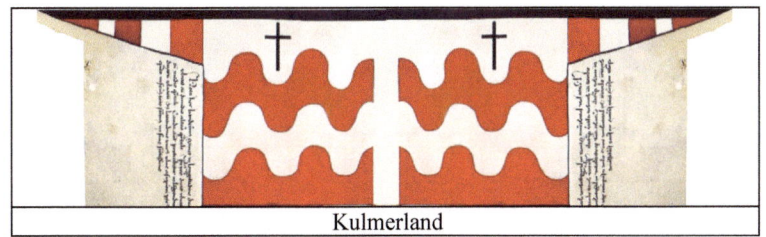
Kulmerland

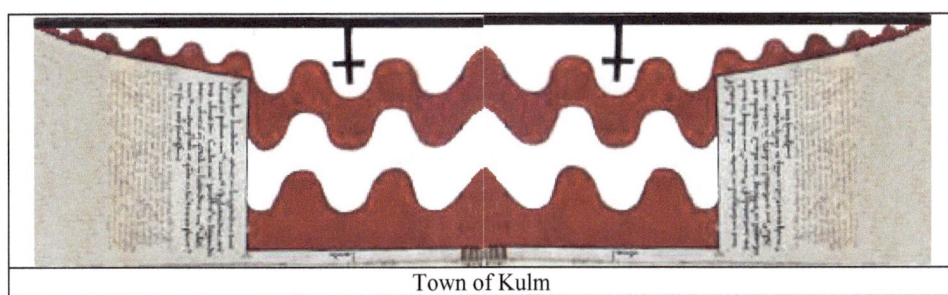
Town of Kulm

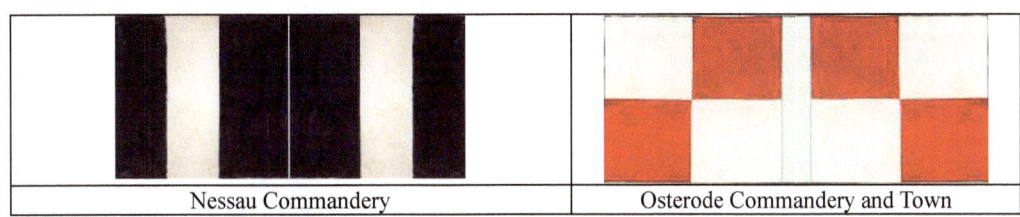

| Nessau Commandery | Osterode Commandery and Town |

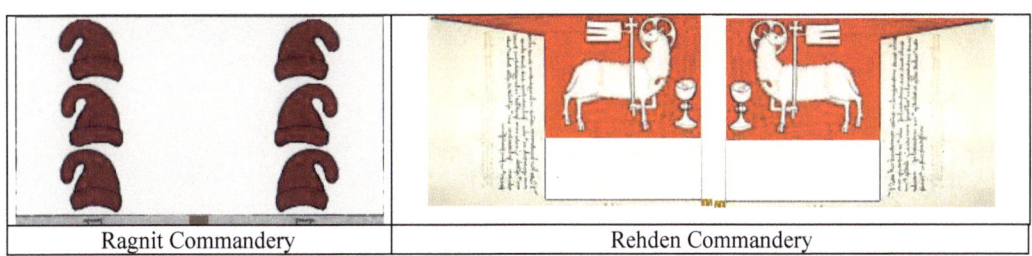

| Ragnit Commandery | Rehden Commandery |

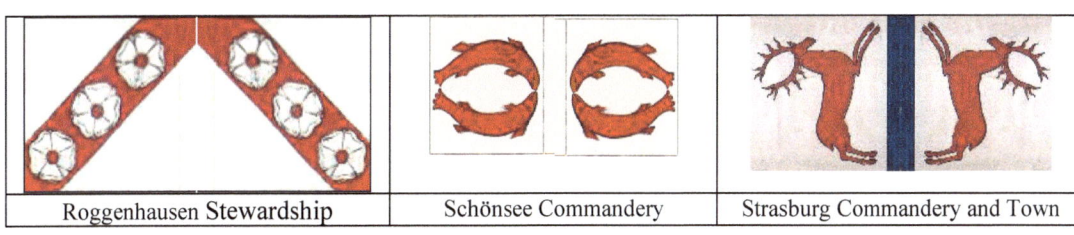

| Roggenhausen Stewardship | Schönsee Commandery | Strasburg Commandery and Town |

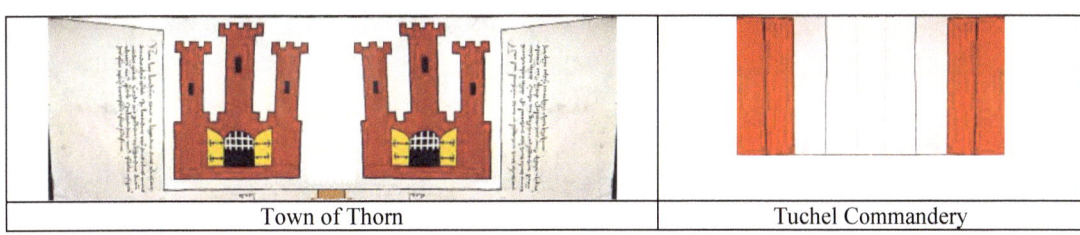

| Town of Thorn | Tuchel Commandery |

83

Allies and Mercenaries

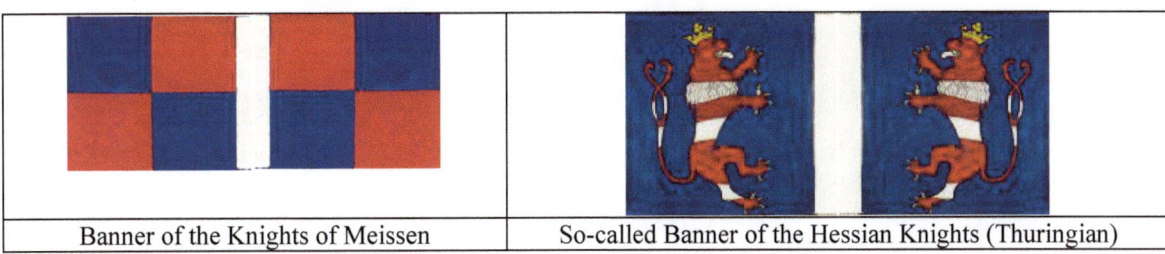

Banner of the Knights of Meissen	So-called Banner of the Hessian Knights (Thuringian)

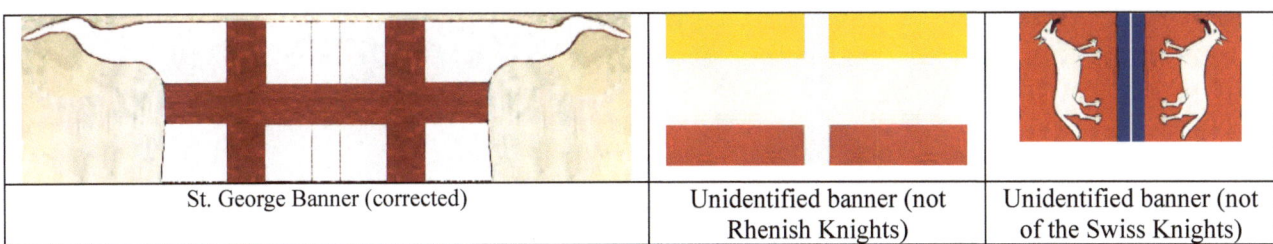

St. George Banner (corrected)	Unidentified banner (not Rhenish Knights)	Unidentified banner (not of the Swiss Knights)

An archers' flag (so-called "Banner of the Westphalian Knights")	An archer's flag (so-called "Banner of the Swabian Knights")	An archers' flag (misidentified as Mewe Commandery)

Teutonic Order Banners Captured by Poland at the Battles of Koronowa in 1410 and at Dąbki/Nakel in 1431

Commandery Schwetz*	Teutonic Order in Livonia (after Vossberg) captured by the Poles in 1431
Ascheraden and Dünaburg Commandery 1431	Fellin Commandery (and Kokenhausen Stewardship) 1431

*The banner of the Schwetz Commandery was said to be captured at Koronowa (German: *Polnisch Krone*) on October 10th, 1410 but the *Komtur* of Schwetz, Heinrich von Plauen, did not take part in this battle, so it is suspected that it could have been the banner of the Steward of the Neumark, Michael Küchmeister, who was taken prisoner.

Annex 2: Teutonic Order Banners Produced by Hans Müller (Erfurt) 1930-1938
(German designations according to his catalog)

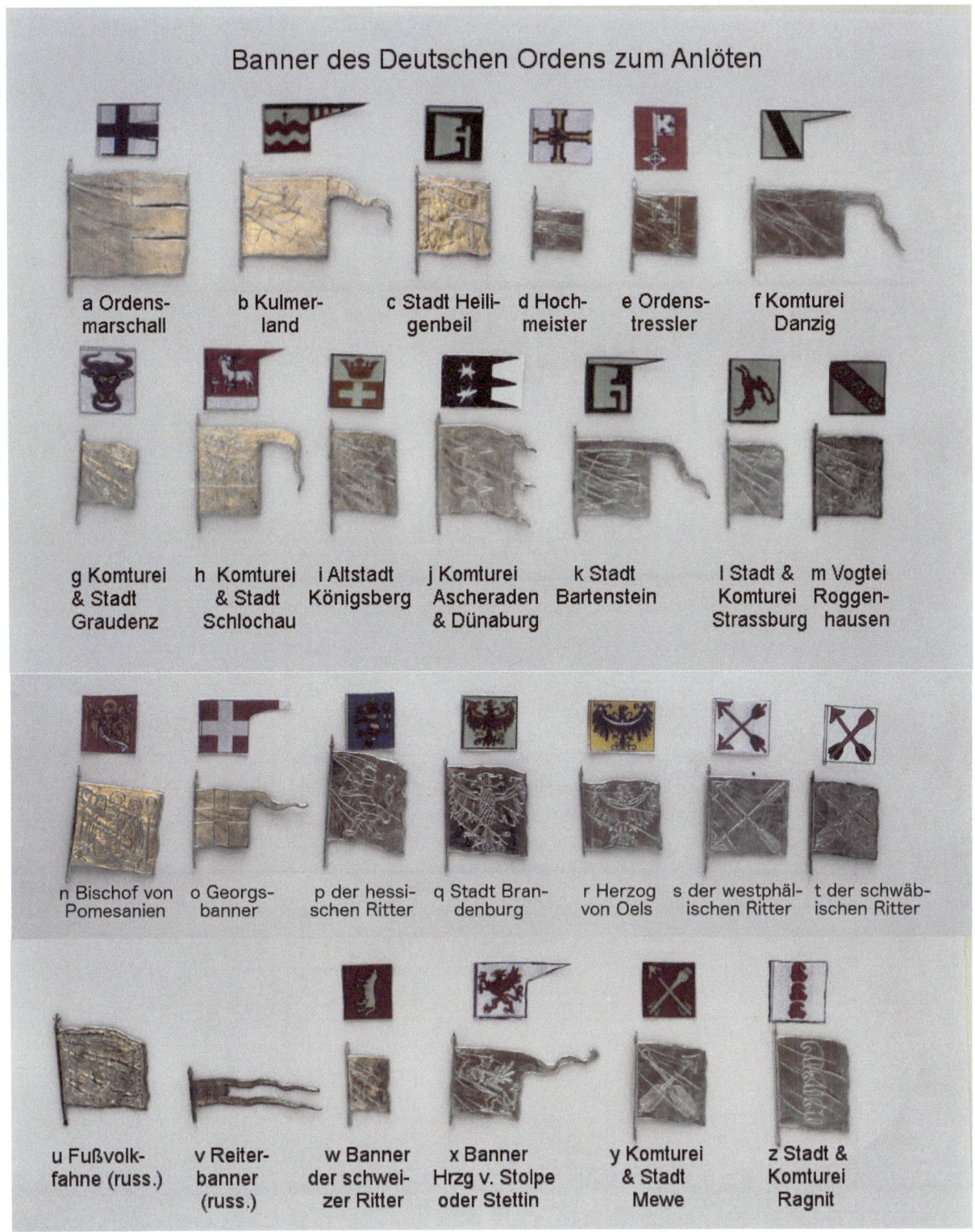

Annex 3: Teutonic Order Banners from the *Banderia Prutenorum*, etc.

The banner's (division's) geographic origin as shown on the map on page 9. Unless otherwise noted, the seat/castle is in Poland; sites listed in Russia (RUS) are in its Kaliningrad *Oblast* (Province).

Banner	Current Name	Page
Grand Master's "Great Banner" (at Marienburg)	Malbork	13-14
Grand Master's "*Rennfahne*" (at Marienburg)	Malbork	14
Grand Commander (*Großkomtur*) (at Stuhm)	Sztum	15
Marshal of the Order (*Ordensmarschall*) (Königsberg)	Kaliningrad, RUS	15
Order's Treasurer (*Ordenstressler*) (at Marienburg)	Malbork	16
Senior Draper (*Oberster Trapier*) (at Christburg)	Dzierzgon,	20
Senior Hospitaller (*Oberster Spittler*) (at Elbing)	Elbląg	17
Bishopric and Bishop of Culm (*Kulm*) - probably the Bishopric and Bishop of Sambia (*Samland*) (at Fischhausen)	Chełmno	17
Bishopric of Warmia (*Ermeland*) alias Town of Heilsberg – see Rheden Commandery	Lidzbark Warmiński	18
Bishop of Pomesania (at Riesenburg)	Prabuty	19
Bishopric and Bishop of Sambia (*Samland*) (at Fischhausen)	Primorsk, RUS	19
St. George Banner (*Georgsbanner*)	N/A	20
Duke of Oels	Oleśnica	23
Duke of Stettin (also mis-identified as the Duke of Stolpe)	Pomorze Szczecińskie	24
Allenstein, Town of (so-called Banner of the Town of Groß-Holsten)	Olsztyn	35
Althaus Commandery	Starogród (Chełmiński)	25
Balga, Commandery and Town of	Weseloe, RUS	26
Barten Trustee's Office (hypothetical)	Barciany	42
Bartenstein Town of	Barciany	26
Birgelau Commandery (hypothetical)	Bierzgłowo	42
Bischöflich Papau Commandery (hypothetical)	Papowo Biskupie	42
Brandenburg, Town of	Uszakowe, RUS	27
Brathien, Stewardship (*Vogtei*) of and of Neustadt	Bratian	28
Braunsberg, Town of	Braniewo	28
Bütow Trustee's Office (hypothetical)	Bytów	43
Danzig Commandery (Commander's banner)	Gdańsk	30
Danzig (Vice Commander's banner)	Gdańsk	30
Danzig Commandery and Town	Gdańsk	30
Dirschau Stewardship and Town	Tczew	31
Elbing, Town and Commandery (Commander's)	Elbląg	32
Elbing, Town and Commandery (Vice-Commander's)	Elbląg	32
Elbing Burghers	Elbląg	32
Engelsburg Commandery	Pokrzywno	31
Gollub Commandery (hypothetical)	Golub	43
Graudenz, Commandery and Town of	Grudziądz	33
Heiligenbeil, Town of	Mamonovo, RUS	33
"Hessian Knights" so-called "Banner of the" (probably a Thuringian levy's flag)	N/A	22
Insterburg Trustee's Office (hypothetical)	Chernyakhovsk, RUS	43
Königsberg	Kaliningrad, RUS	34
Königsberg, Altstadt (Old Town)	Kaliningrad, RUS	34
Kulm, Town of (or Culm)	Chełmno	28
Kulmerland (or Culmerland)	Chełmno	29
Lessen or Leske, Stewardship of	Łasin or Leszken	34
Meissen, the Knights of	Meissen, Germany	22

Banner	Current Name	Page
Memel Commandery (hypothetical)	Klaipéda, Lithuania	43
Mewe, so-called "Banner of the Commandery the Castle and the Town of Gnyew"	Gniew	36
Nessau Commandery	Nieszawa	35
Ortelsburg Commandery (see Schwetz Commandery)	Szczytno	36
Osterode Commandery	Ostróda	37
Preußisch-Mark (Pruschenmarkt or Preuschmark) Commandery	Przezmark	44
Ragnit Commandery	Neman, RUS	37
Rheden Commandery (see Schlochau Commandery and Town)	Radzyń Chełmiński	39
Rhein Commandery (hypothetical)	Ryn	44
"Rhine and Germany" so-called "Knights from"(a common battle standard)	N/A	23
Roggenhausen Stewardship & Town	Rogóźno	38
Schlochau, Commandery and Town of	Człuchów	39
Schönsee, Commandery and Town of	Kowalewo	40
Schwetz or Ortelsburg Commandery	Szczytno	36
Soldau Commandery (hypothetical)	Działdowo	44
Strasburg, Town and Commandery of	Brodnica	40
"Swabian Knights", so-called "Banner of the "(an archers' flag)	N/A	21
"Swiss Knights", so-called "Banner of the" (more likely a second Balga Commandery flag)	N/A	24
Thorn, Town of	Toruń	40
Tuchel, Commandery and Town of	Tuchola	41
Wenzlau Trustee's Office (hypothetical)	Unisław	44
"Westphalian Knights", so-called "Banner of the" (an archers' flag)	N/A	25
Master of the Livonian Order (1431) (at Wenden)	Cēsis, EST	45
Ascheraden and Dünaburg Commandery (1431)	Aizkraukle, Latvia	46
Fellin Commandery (1431)	Viljandi, Latvia	46
Kokenhausen, Stewardship of (1431)	Koknese, Latvia	46-47

Marienburg Fortress, Malbork, Poland, July 2013 – Photo: R. Sanders

Annex 4: Günter Fricke's Map:
The Polish-Lithuanian Kingdom in 1400

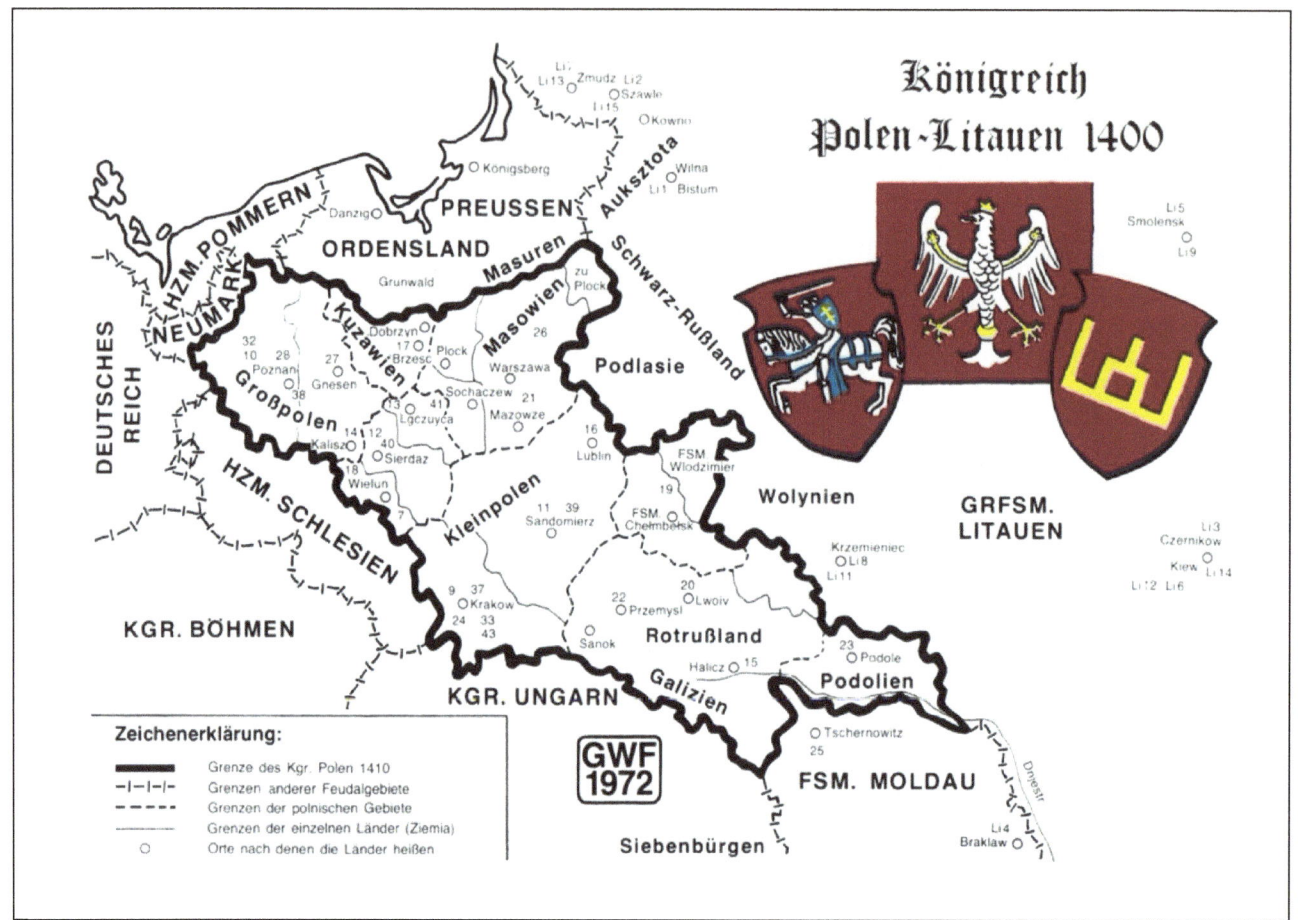

The Polish-Lithuanian Kingdom in 1400 by Günter W. Fricke (1972)
The numbers refer to the locations where the coats of arms appeared

Legend:

    ___ Border of the Polish Kingdom in 1410
-|-|-|- Borders of other feudal states
- - - - Borders of Polish regions
    _____ Borders of the individual states (*Ziemia*)
    o Places for which the states are named

Place names in alphabetical order other than towns/cities:

DEUTSCHES REICH = German Empire	Kleinpolen = Lesser Poland/Małopolska
FSM … = Principality of …	Kujawien = Kuyavia
FSM MOLDAU = Principality of Moldavia	Masowien = Mazovia
GRFSM. LITAUEN = Grand Duchy of Lithuania	Masuren = Masuria
Großpolen = Greater Poland/Wielkopolska	ORDENSLAND PREUSSEN = Order's State of Prussia
HZM. POMMERN = Duchy of Pomerania	
HZM SCHLESIEN = Duchy of Silesia	Rotrußland = Red Ruthenia
KGR. BÖHMEN = Kingdom of Bohemia	Schwarz-Rußland = Black Russia (Belorus)
KGR. UNGARN = Hungarian Kingdom	Siebenbürgen = Transylvania

Annex 5: Polish Banners from the *Banderia Apud Grunwald I* Book

The flags of the 37th, 39th, 41st and 45th Banners (divisions) are not listed in the normal order in the "Fig." (image) column because in Andrzej Klein's book they are paired with a lower numbered flag. Lithuanian and Ruthenian flags that only appear in Günter W. Fricke's articles are integrated without numbers. The map number corresponds to the banner's (division's) geographic origin as shown on the map in Annex 4 (page 89).

Flag Fig.	Banner	Banner No.	Map No. (per Fricke)	Page
1a	"Great" Kraków Banner	1	9	54
1b	Royal Pennon / Bodyguard	1	9	54
2	Banner "Gończa" / "Flying Band"	2	-	55
3	Nadworna / "Court Band" Banner	3	-	55
4	Banner of St. George	4	-	56
5	Banner of the Poznań District	5	10	56
6	Banner of the Sandomierz District	6	11	56-57
7	Banner of the Kalicz District	7	14	57
8	Banner of the Sieradz District	8	12	57
9	Banner of the Lublin District	9	16	58
10	Banner of the Łęczyca District	10	13	58
11	Banner of the Kuyavian District	11	17	58
12	Banner of the District of Lwów	12	20	59
13	Banner of the Wieluń District	13	7 & 18	59
*	Banner of the Wieluń District (with a lamb)		7 & 18	59
*	Banner of the Wieluń District (with horns)		7 & 18	59
14	Banner of the Przemyśl District	14	22	60
15	Banner of the Dobrzyń District	15	17	60
16	Banner of the Chelm District	16	19	61

Polish Banners from the book *Banderia Apud Grunwald I*, p. 123

Polish Banners from the *Banderia Apud Grunwald I* Book

Flag Fig.	Banner	Banner No.	Map No.	Page
17-19	Banner of the Podolien District	17, 18 & 19	23	61
20	Banner of the Halicz District	20	15	62
21, 22	Banner of Duke Siemowit of Mazovia / Mazovian District	21 & 22	21	62
23	Banner of Duke Janusz of Mazovia	23	21	62-63
24	Banner of Archduke Mikołaj Kurowski of Gniezno	24	27	63
45	Banner of Mikołaj Kmita of Wiśnicz	45	48	68
25	Banner of Wojciech Jastrzębiec, Bishop of Poznań	25	28	63
26	Banner of Krystyn of Ostrów, the Castellan of Kraków	26	33	63
27	Banner of Jan of Tarnów, the Voivode of Kraków	27	37	64
37	Banner of Wincentego of Granowo, the Castellan of Nakło and Starosta-General of Greater Poland/Wielkopolska	37	32	64, 66
39	Banner of Spytko of Jarosław and Tarnów, the Starost-General of the Ruthenians and Voivode of Sandomierz	39	39	64
28	Banner of Sędziwój of Ostroróg (symbol of the Nałęcz Clan)	28	38	64
41	Banner of Dobrogost Świdwa of Szamotuli (symbol of the Nałęcz Clan)	41	-	64
29	Banner of Mikołaj of Michałów, the Voivode of Sandomierz (with "Roja" coat of arms)	29	39	64
30	Banner of the Jakub of Koniecpol, the Voivode of Sieradz	30	40	64
31	Banner Jan of Obichów, the Castellan of Śrem	31	-	65
32	Banner of Jan Ligęza of Bobrek, the Voivode and *Starosta* of Łęczyca	32	41	65
33	Banner of Andrzej of Tęczyn, the Castellan of Wojnice	33	-	65
34	Banner of Zbigniew of Brzezie, Marshal of the Polish Kingdom	34	31	66
35	Banner Pjotr Szafraniec of Pieskowa Skała, the Chamberlain of Kraków	35	43	66
36	Banner of Klemens of Moskorzew, the Castellan of Wiślica	36	-	66
38	Banner of Dobiesław of Oleśnica	38	44	66-67
40, 40a	Banner of Marcin of Sławsko ("Zaremba" device from *Armorial of the Golden Fleece*)	40	45	67

Polish Banners from the book *Banderia Apud Grunwald I*, p. 125

Polish Banners from the *Banderia Apud Grunwald I* Book

Flag Fig.	Banner	Banner No.	Map No.	Page
42	Banner of Krystyn of Koziegłowy, the Castellan of Sącz	42	36	67
43	Banner of Jan Mężyk of Dąbrowa	43	46	68
44	Banner of Mikołaj Trąba of Wiślicz	44		68
46	Banner of the Knight Brothers of the Griffin (*Gryf*)	46	30	68
47	Banner of the Knights of Zaklika of Korzkiew	47	29	69
48	Banner of the Knight Brothers of the Koźle Rogi Clan (Jelita)	48	42	69
49	Banner of the Moravian Baron Jan of Jičín (Jana z Jičína)	49	6	69
50	Banner of Gniewosz of Dalewice	50	-	70
	Banner of the Bishop of Kraków		9	70
	Banner of the Prince of Mazovia		21	70-71
51	Banner of the Lithuanian Duke Zygmunt, Son of Korybut (Korybutowicza)	51	-	72
A	Lithuanian Banner (Pogoń horseman)		-	72
	Auksztota (after G. W. Fricke, L I) / Gates or Pillars of the Gediminids		Li 1	72
	Braclaw (after G. W. Fricke, L XI)		Li 4	73
	Żmudź (after G. W. Fricke, L IV)		Li 7	73
	Krzemienice (after G. W. Fricke, L IX)		Li 8	73
	Samogetia			73
B	Smolensk		Li 5	74
	Smolensk (after Schenzle's Li 5),		Li 5	74
	Kiev (after G. W. Fricke's L VII)		Li 14	75
	Czernikow (Chernihiv) (after G. W. Fricke's L X)		Li 3	75
C	Tartars		-	76

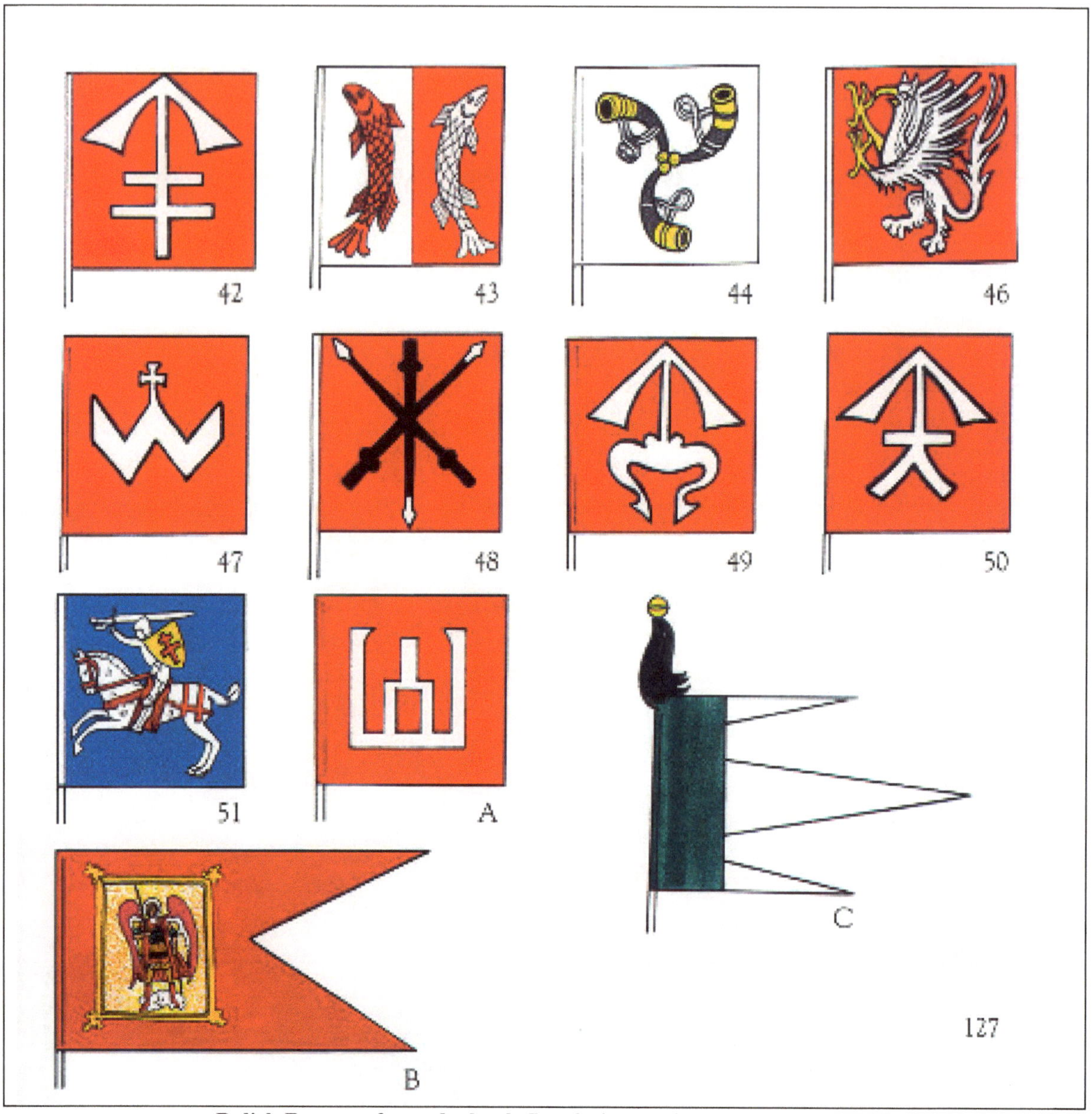

Polish Banners from the book *Banderia Apud Grunwald I*, p. 127

Annex 6: Polish-Lithuanian Flags by Erwin Ortmann

Erwin Ortmann (†) also created color illustrations of Polish flags from Tannenberg, including four which were variations of those by Günter W. Fricke, shown in Annex 7.

Selected Polish-Lithuanian flags created by Erwin Ortmann (†) that do not appear in either Heinz Schenzle's or Andrzej Klein's works

96

Annex 7: Günter W. Fricke's
Polish-Lithuanian Banners and Those of Their Allies at Tannenberg

Polish-Lithuanian Banners at Tannenberg, plate 1 (1979?) by Günter Fricke, 1985

Polish-Lithuanian Banners at Tannenberg, plate 2 (1979) by Günter Fricke, 1985

Polish-Lithuanian Banners at Tannenberg, plate 3 (1979) by Günter Fricke, 1985

Polish-Lithuanian Banners at Tannenberg, plate 4 (1980) by Günter Fricke, 1985

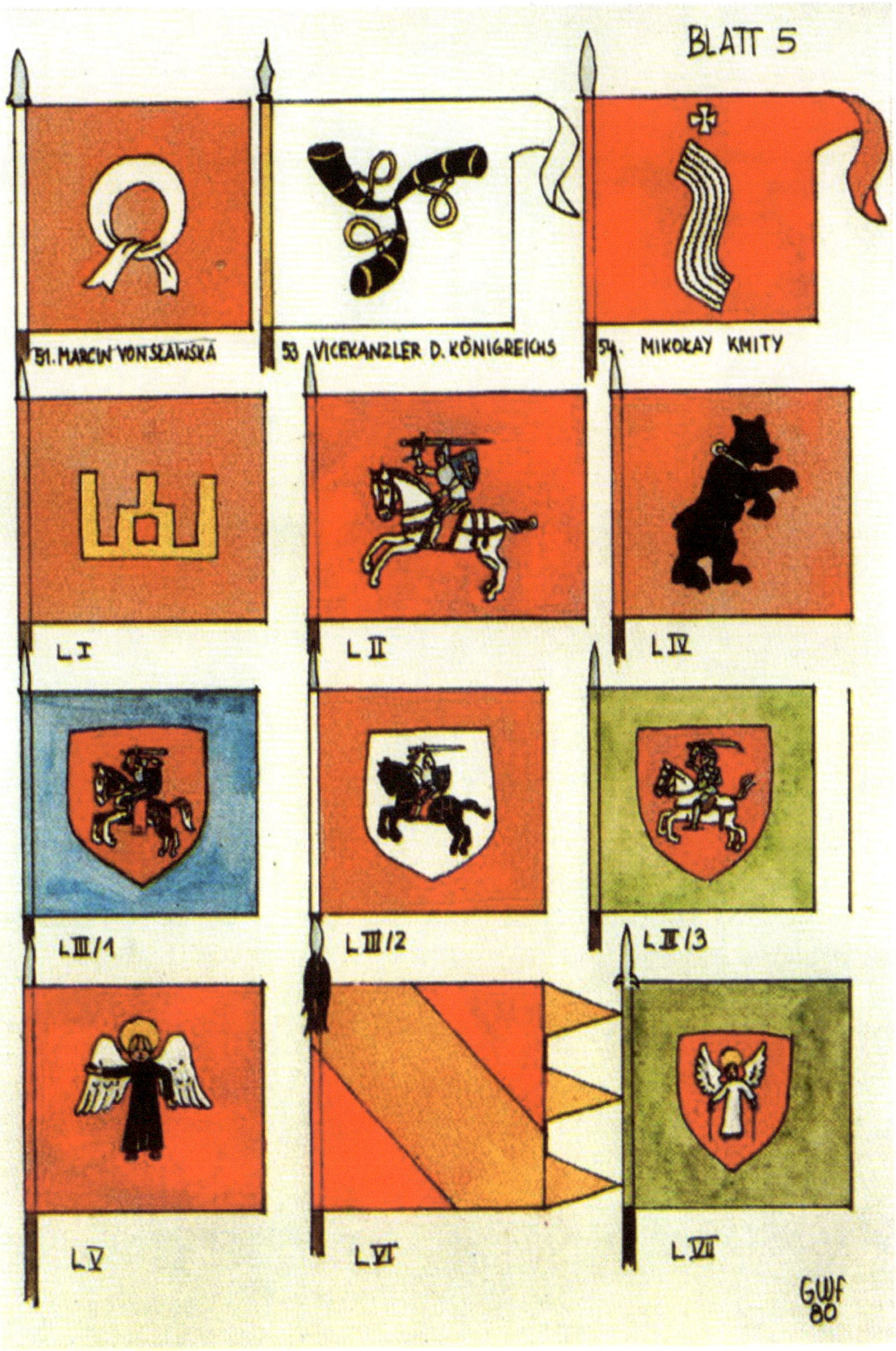

Polish-Lithuanian Banners at Tannenberg, plate 5 (1980) by Günter Fricke, 1985

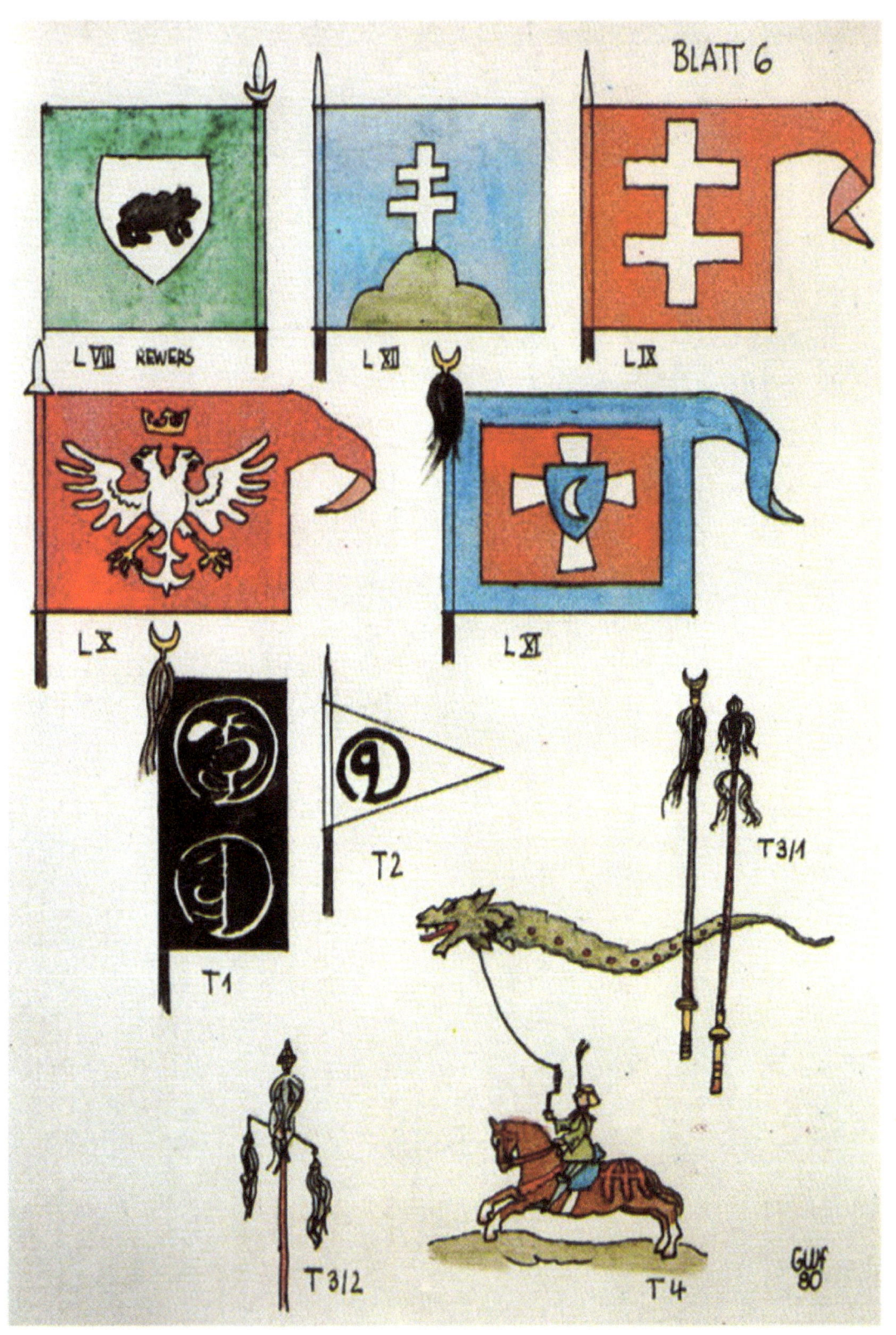

Polish-Lithuanian Banners at Tannenberg, plate 6 (1980) by Günter Fricke, 1985

Annex 8: Polish-Lithuanian Coats of Arms after Günter W. Fricke

1. Awadaniec + Habdank
2. Bogoria
3. Ciołek
4. Dębno
5. Dołęga
6. Janina
7. Jastrzębiec
8. Jelitta
9. Junosza
10. Korczak
11. Leliwa
12. Lis
13. Lubicz
14. Łabędzie
15. Łódźia
16. Nałęcz
17. Nowina
18. Odrowaz
19. Ogończyk
20. Pilawa

Annex 9: Polish Coats of Arms from the *Armorial of Knights of the Golden Fleece* (from the book *Banderia Apud Grunwald I*)

Polskie herby w *Herbarzu Złotego Runa*.
Polish coats of arms in the *Golden Fleece Armorial*.

Polish Coats of Arms from the *Armorial of Knights of the Golden Fleece* from *Banderia Apud Grunwald I*, p. 7

Polskie herby w *Herbarzu Złotego Runa*.
Polish coats of arms in the *Golden Fleece Armorial*.

8

Polish Coats of Arms from the *Armorial of Knights of the Golden Fleece* from *Banderia Apud Grunwald I*, p. 8

Polskie herby w *Herbarzu Złotego Runa*.
Polish coats of arms in the *Golden Fleece Armorial*.

Polish Coats of Arms from the *Armorial of Knights of the Golden Fleece* from *Banderia Apud Grunwald I*, p. 10

Annex 10: Polish Coats of Arms from the *Codex Bergshammar*

The *Codex Bergshammar* armorial resides in the Swedish National Archives. According to the Heraldica Nova website, "This fascinating parchment manuscript contains over 3,000 coats of arms, displaying first European high nobles by rank followed by nobles by political entities.... It is dated to the mid-15th century and it is assumed to have been made in the Southern Netherlands. Regrettably, nothing is known about the maker and context of use."

Source: Bergshammarsamlingen: https://commons.wikimedia.org/wiki/ File:Bergshammarsamlingen

Source: Bergshammarsamlingen: https://commons.wikimedia.org/wiki/ File:Bergshammarsamlingen

Source: Bergshammarsamlingen: https://commons.wikimedia.org/wiki/ File:Bergshammarsamlingen

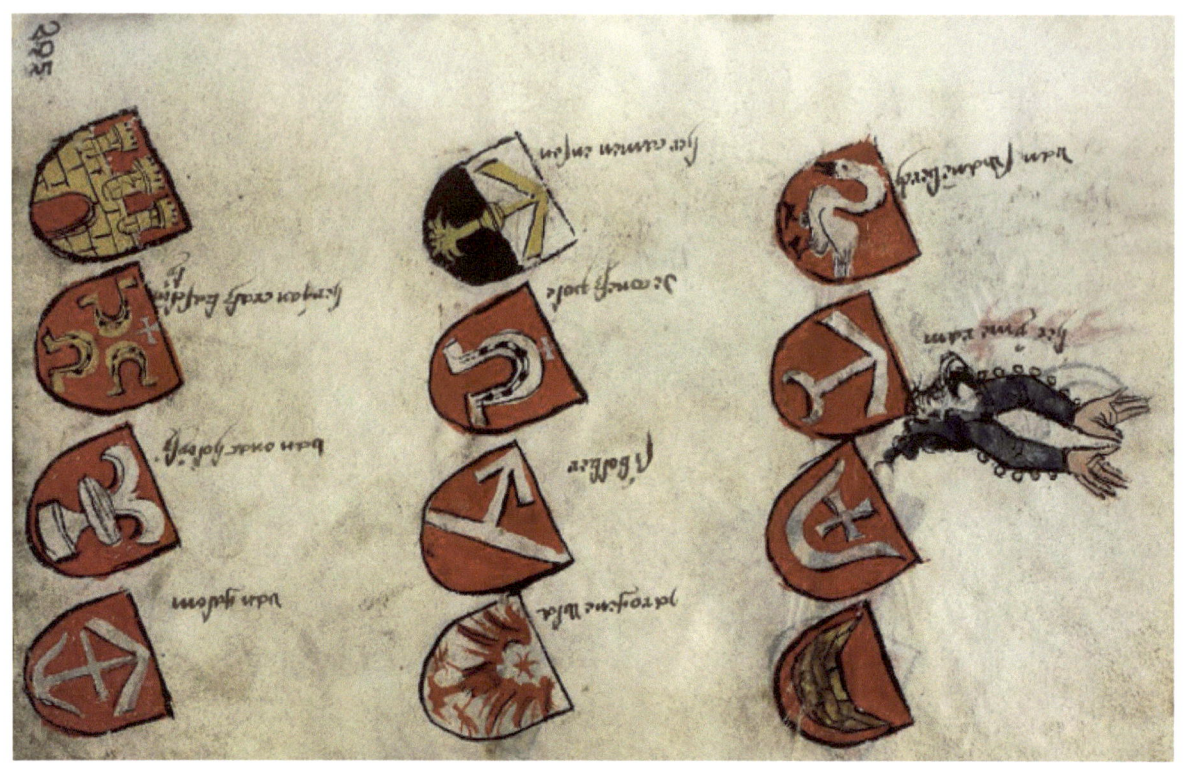

Source: Bergshammarsamlingen: https://commons.wikimedia.org/wiki/ File:Bergshammarsamlingen

Annex 11: Contributors

(in Germany unless otherwise noted)

Bähr, Frank, Schkeuditz
Bistulfi, Gianpaolo, Milano, Italy
Blawatt, Peter, Saalfeld
Brümmer, Dr. Thomas, Halle/Saale
Büche, Wolfgang (Halle/Saale)
Bunde, Peter, Herzogenrath
Ewert, Dirk, Bochum
Friedrich, Wolfgang, Rackwitz, Saxony
Fuhrmann, Rolf, Wilhemshaven
Grimm, Bernt, Borlange, Sweden
Levec, Manfred, Sindelfingen
Meissner, Horst, Drößnitz
Sanders, Rick, Candler, North Carolina, USA
Schlager, Gernot, Leonding, Austria
Schmidt, Henry, Berlin
Schwahn, Hans, Dessau and Moscow, Russia
Schwarz, Jürgen, Stuttgart
Sennebo, Jan, Helsingborg, Sweden
Wachter, Stefan, Kulmbach

Painters of figures from the other individuals' collections:
    Hensel, Jörg, Leipzig, from Wolfgang Friedrich's collection
    Kröbel, Maurice, Dessau, from Hans Schwahn's collection

Polish Banners produced by Hans Jörg Stoll; painted & photo by R. Sanders

Annex 12: Sources and Recommended Reading

Archivrath, Dr. Joachim, *Das Marienburger Treßlerbuch der Jahre 1399-1409* [Marienburg Treasurer's Book for 1399-1409] (Königsberg: Thomas & Ottermann, 1896).

Bergshammarsamlingen (*Codex Bergshammar*): https://commons.wikimedia.org/wiki/File:Bergshammarsamlingen

Długosz, Jan, *Annales seu cronicae incliti regni Poloniae* (1455-1480).

-- *Banderia Prutenorum* (1455-1480).

-- *Insignia seu clenodia Regni Poloniae* (1455-1480).

Ekdahl, Sven, *Die "Banderia Prutenorum" des Jan Długosz - eine Quelle zur Schlacht bei Tannenberg 1410 : Untersuchungen zu Aufbau, Entstehung u. Quellenwert der Handschrifts* (Gottingen: Vandenhoeck und Ruprecht, 1976).

Engel, Bernhard (ed.), *Die mittelalterlichen Siegel des Thorner Rathsarchiv*, vol. 1. (Danzig: Saunier, 1902).

Fricke, Günter W., "Die Banner des polnisch-litauischen Heeres in der Schlacht bei Grunwald 1410", *Arbeitsmaterial Kulturgeschichtliche Zinnfiguren* (Kulturbund der DDR, 1985), pp. 14-19.

Fuhrmann, Rolf; Isclt, Gerald, *Tannenberg 1410: Die Niederlage des Deutschen Ritterordens; Die Belagerung der Marienburg 1410* (Berlin: Heere & Waffen 7, Zeughaus Verlag, 2008, ISBN 978-3-938447-37-6).

Fuhrmann, Rolf, *Der Deutschorden – von Akkon bis zum Baltikum. Die Armee 1198-1420* (Berlin: Zeughaus, 4th edition, 2017).

Iselt, Gerald / Fuhrmann, Rolf, *Tannenberg 1410 – Die Belagerung der Marienburg 1410* (Berlin: Zeughaus, 2008).

Jeziorowski, Tadeusz, *Grunwald 1410 w artystycznej rekonstrukcji historycznej* [Grunwald 1410 with artistically reconstructed history], (Poznań: Muzeum Narodowe w Poznaniu, 2010), ISBN 978-83-89053-84-8.

Klein, Andrzej / Sekunda, Nicholas / Czernielewski, Konrad A., *Banderia Apud Grunwald I: Chorągwie polskie pod Grunwaldem – Polish Banners at Grunwald* (Łódź: Alexander, 2000).

Klein, Andrzej / Nowakowski, Piotr; *Banderia Apud Grunwald II: Chorągwie krzyżackie pod Grunwaldem – Teutonic Banners at Grunwald* (Łódź: Alexander, 2000).

Klietmann, "Kleiner Beitrag zur Geschichte der Schlacht bei Tannenberg im Jahre 1410", *Die Zinnfigur*, 2/82, pp. 45-46, 1982.

Matwijewicz, Tadeusz, Geneza Herbu Gminy Człuchów krótka historia człuchowskich herbów (https://ugczluchow.pl/files/file/Grzesiek/Geneza.pdf).

Michael, Maurice, *The Annals of Jan Długosz: Annales seu cronicae incliti regni Poloniae* (Chichester, West Sussex: IM Publications, 1997).

Paravicini, Werner, *Die Preußenreisen des Europäischen Adels. Teil 2* [The Prussian Journeys of the European Nobility, Part 2] (Sigmaringen: Jan Thorbecke Verlag, 1995).

Sanders, Richard, "Teutonic Order Banners Captured by Poles at the 1410 Battle of Tannenberg from *Banderia Prutenorum*", National Capital Model Soldier Society, *The Dispatch,* July 2021, pp. 33-40.

Sarnecki, Witold; Nicolle, David, Medieval Polish Armies 966-1500 (Oxford: Osprey Publishing, Men-at-Arms Series No. 445, 2008), ISBN 978-1-84603-014-7.

Schenzle, Heinz, Schlacht bei Tannenberg 1410 Kampf um die Marienburg (Schwaig b. Nürnberg: Verlag E. Kästner, 1982).

Turnbull, Stephen, *Tannenberg 1410: Disaster for the Teutonic Knights* (Botley, Oxford: Osprey, 2003).

Turnbull, Stephen, *Crusader Castles of the Teutonic Knights (1): The red-brick castles of Prussia 1230-1466* (Botley, Oxford: Osprey, 2003 – Fortress 11).

Turnbull, Stephen, *Crusader Castles of the Teutonic Knights (2): The stone castles of Latvia and Estonia 1185-1560* (Botley, Oxford: Osprey, 2004 – Fortress 19).

Vossberg, F. A. (ed.), *Banderia Prutenorum oder die Fahnen des Deutschen Ordens und seiner Verbündeten, welche in Schlachten und Gefechten des 15. Jahrhunderts eine Beute der Polen wurden*, (Berlin: C. Feister, 1849).

Vossberg, F. A., *Geschichte der preußischen Münzen und Siegel bis zum Ende der Herrschaft des Deutschen Ordens* (Berlin: Fincke, 1842) (shown in endnotes as "*Münzen und Siegel*").

Ziesemer, W., *Das Große Ämterbuch des Deutschen Ordens* [The Great Official Register of the Teutonic Order] (Danzig: A. W. Kafemann, 1921).

Endnotes

[1] Fuhrmann, Rolf, *Der Deutschorden – von Akkon bis zum Baltikum. Die Armee 1198-1420*, (Berlin: Zeughaus, 4th ed., 2017), p. 28.

[2] Archivrath, Dr. Joachim, *Das Marienburger Treßlerbuch der Jahre 1399-1409* [Marienburg Treasurer's Book for 1399-1409] (Königsberg: Thomas & Ottermann, 1896).

[3] Tautorat, Hans-Georg, *Schwarzes Kreuz auf weißem Mantel. Die Kulturleistung des Deutschen Ordens in Preußen* [Black Cross on a White Cloak: The Cultural Accomplishments of the Teutonic Order in Prussia] (Düsseldorf: NWZ Verlag, 1977).

[4] Ibid and Weber, Lothar, *Preußen vor 500 Jahren*, [Prussia 500 Years Ago] (Danzig: T. Bertling Verlag, 1878).

[5] Weber, Lothar, *Preussen vor 500 Jahre* (Danzig: 1878).

[6] Ziesemer, W., *Das Große Ämterbuch des Deutschen Ordens* [The Great Official Register of the Teutonic Order] (Danzig: A. W. Kafemann, 1921).

[7] Ekdahl, Sven, *Die "Banderia Prutenorum" des Jan Długosz Eine Quelle zur Schlacht bei Tannenberg 1410* [The „Banderia Prutenorum" of Jan Długosz – a Source on the Battle of Tannenberg), *pp. 24-25*.

[8] Paravicini, Werner, *Die Preußenreisen des Europäischen Adels. Teil 2* [The Prussian Journeys of the European Nobility, Part 2] (Sigmaringen: Jan Thorbecke Verlag, 1995).

[9] Ibid.

[10] https://de.wikipedia.org/wiki/Banderia_Prutenorum

[11] Vossberg, F. A. (ed.), *Banderia Prutenorum oder die Fahnen des Deutschen Ordens und seiner Verbündeten, welche in Schlachten und Gefechten des 15. Jahrhunderts eine Beute der Polen wurden* (Berlin, 1849).

[12] For the Kästner figures with engraved flags, see https://www.schmalkalder-zinnfiguren.de/30-mm-flachfiguren/mittelalter/tannenberg-grunwald-1410/deutscher-orden/banner-des-deutschen-orden-in-der-schlacht-bei-tannenberg-1410/. These Teutonic Order figures were designed by Karl Heinrichs and engraved by Regina Sonntag and Karl-Werner Rieger and are identified here with their ordering numbers TB 02 through TB 41.

[13] Archivrath, Dr. Joachim, *Das Marienburger Treßlerbuch der Jahre 1399-1409*.

[14] Vossberg, *Banderia Prutenorum*, p. 15; plate I, No. 2.

[15] Ibid., p. 16; plate I, No. 3.

[16] Vossberg, F. A., *Münzen und Siegel*, plate VIII, No. 9.

[17] Fuhrmann, Rolf, *Der Deutschorden*, p. 40.

[18] Vossberg, *Banderia Prutenorum*, p. 17; plate I, No. 4.

[19] Perlbach, Max, *Die Statuten des Deutschen Ordens nach den ältesten Handschriften* [The Statutes of the Teutonic Order According to the Oldest Hand-Written Manuscripts] (Halle: Gebrüder Bornträger, 1890).

[20] Vossberg, *Banderia Prutenorum*, p. 18; and plate I, No. 5.

[21] Ibid., p. 18; and plate II, No. 6.

[22] Ibid., pp. 19-20; and plate II, No. 7 and plate VII, No. 47.

[23] Ibid., pp. 20-21; and plate II, No. 8.

[24] Ibid., p. 22; and plate II No. 9

[25] Ekdahl, *Banderia Prutenorum*, p. 194, footnote 2.

[26] Vossberg, *Banderia Prutenorum*, pp. 22-24; and plate II, No. 10.

[27] Michael, Maurice, The Annales of Jan Długosz, pp. 399-400.

[28] Vossberg, *Banderia Prutenorum*, pp. 24-25; and plate II, No. 11.

[29] Ibid., p. 25.

[30] Paravicini, *Die Preußenreisen des Europäischen Adels*.

[31] Ibid.

[32] Ibid.

[33] Ekdahl, Sven, *Banderia Prutenorum"*, pp. 90-93.

[34] Vossberg, *Banderia Prutenorum*, pp. 25-26; and plate III, No. 12.

[35] Ibid., pp. 26-27; and plate III, No. 13.

[36] Ibid., p. 27; and plate III, No. 14.

[37] Ibid., pp. 27-28; and plate III, No. 15.

[38] Ibid., pp. 28-29; and plate III, No. 16.

[39] Ibid., pp. 29-30; and plate III, No. 17.

[40] Ibid., pp. 30-31; and plate III, No. 18.

[41] Ibid., pp. 31-32; and plate III, No. 19.

[42] Ibid., p. 32; and plate IV, No. 20.
[43] Ibid., pp. 32-33; and plate IV, No. 21.
[44] Michael, Maurice, *The Annales*, p. 395.
[45] Vossberg, *Banderia Prutenorum*, p. 33; and plate IV, No. 22.
[46] Ekdahl, *Banderia Prutenorum*, p. 230 footnotes 2 and 3; p. 270, footnotes 2 and 3.
[47] Vossberg, *Banderia Prutenorum*, pp. 33-35, and plate IV, No. 23.
[48] Ekdahl, *Banderia Prutenorum*, p. 248 footnote 2.
[49] Michael, Maurice, *The Annales*, p. 395.
[50] Vossberg, *Banderia Prutenorum*, p. 36, and plate IV, No. 24.
[51] Ibid., p. 36; and plate IV, No. 25.
[52] Ibid., pp. 37-38; and plate IV, No. 26.
[53] Ibid., pp. 38-39; and plate IV, No. 27).
[54] Ibid., pp. 39-40; and plate V, No. 29.
[55] Ibid., p. 40; and plate V No. 29.
[56] Ibid., p. 41, according to Vossberg, "*Because King Wladislaw of Poland, when he was in Danzig in 1457, along with granting other boons to the city, added the right among others to add a gold royal crown to the two Order's crosses ..., so this sheds light on the fact that this had been spoken of having happened a long time before, that when in 1448 Długosz handed over the Banderia manuscript to the Krakow Cathedral, either he later revised it, or it had already be expanded with further additions.*"
[57] Ibid., pp. 40-41; and plate V No. 28.
[58] Ibid., pp. 42-43; and plate V, No. 30.
[59] Ibid., p. 44; and plate V No. 34.
[60] Ekdahl, *Banderia Prutenorum*, p. 208, footnote 3.
[61] Ibid., p. 43; and plate V, No. 32.
[62] Ibid., pp. 42-43; and plate V, No. 31.
[63] Ibid., pp. 43-44; and plate V, No. 33.
[64] Vossberg, *Banderia Prutenorum*, pp. 44-46; and plate V, No. 35.
[65] Ibid., p. 46; and plate VI, No. 36.
[66] Ekdahl, *Banderia Prutenorum*, p. 270, footnotes 2 and 3.
[67] Vossberg, *Banderia Prutenorum*, pp. 46-47; and plate VI, No. 37.
[68] Ekdahl, *Banderia Prutenorum*, p. 190, footnote 2.
[69] Vossberg, *Banderia Prutenorum*, pp. 47-48; and plate VI, No. 38.
[70] Ibid., p. 48; and plate VI, No. 39.
[71] Ekdahl, *Banderia Prutenorum*, p. 224, footnote 2.
[72] Vossberg, *Banderia Prutenorum*, pp. 48-49; and plate VI, No. 40.
[73] Vossberg, *Banderia Prutenorum*, pp. 51-52; and plate VI, No. 42.
[74] Ibid., pp. 49-51; and plate VI, No. 41.
[75] Ibid., pp. 51-52; and plate VI, No. 43.
[76] Michael, Maurice, *The Annales*, p. 395.
[77] Ibid., pp. 52-53; and plate VII, No. 44.
[78] Ibid., p. 53; and plate VII, No. 45.
[79] Michael, Maurice, *The Annales*, p. 395.
[80] Vossberg, *Banderia Prutenorum*, pp. 53-55; and plate VII, No. 46.
[81] Eckdahl, *Banderia Prutenorum*, p. 204, footnote 3.
[82] Vossberg, *Geschichte der preußischen Münzen und Siegel*, p. 39; and plate XV.
[83] Engel, Bernhard (ed.), *Die mittelalterlichen Siegel des Thorner Ratsarchivs*, vol I, p. 5; and plate III.
[84] Vossberg, *Banderia*, pp. 55-56; and plate VII, No. 47.
[85] Michael, Maurice, *The Annales*, p. 395.
[86] Tadeusz Matwijewicz, Geneza Herbu Gminy Człuchów krótka historia człuchowskich herbów.
[87] Vossberg, *Münzen und Siegel*, p. 39, plate XV, nos. 35 and 36.
[88] Engel, Bernhard (ed.), *Die mittelalterlichen Siegel des Thorner Rathsarchiv*, vol. 1, p. 5.
[89] Schlochau: (https://ugczluchow.pl/files/file/Grzesiek/Geneza.pdf)
[90] Vossberg, *Banderia*, p. 56; and plate VII, No. 48.
[91] Ibid., pp. 56-57; and plate VII, No. 49.
[92] Michael, Maurice, *The Annales*, p. 395.
[93] Vossberg, *Banderia*, p. 57; and plate VII, No. 50.
[94] Ibid., pp. 58-59; and plate VII, No. 51.
[95] Ekdahl, *Banderia Prutenorum*, p. 96.

[96] Ibid.
[97] Michael, Maurice, *The Annales*, p. 395.
[98] https://www.wikiwand.com/de/Ordensburg_Preußisch_Mark.
[99] Ekdahl, *Banderia Prutenorum*, p. 96, footnote.
[100] Ibid., pp . 63-64 plate VIII, No. 52.
[101] Vossberg, *Banderia*, pp. 62-63; plate VIII, No. 53.
[102] Ibid., pp. 63-64; plate VIII, No. 54.
[103] Ibid., pp. 63-64 plate VIII, No. 54.
[104] Fricke, Günter W., "Die Banner des polnisch-litauischen Heeres in der Schlacht Bei Grunwald 1410", *Arbeitsmaterial Kulturgeschichtliche Zinnfiguren* (Kulturbund der DDR, 1985), p. 14.
[105] Ibid, pp. 14 & 18.
[106] Email correspondence between the author and Wolfgang Büche.
[107] "Bitwa pod Grunwaldem" painted by Stefana Garwatowskiego, source: https://opinie.olsztyn.pl/historia/pomnik-bitwy-grunwaldzkiej-w-uzdowie/#.YyjhdC-B3yU
[108] Men from Samogitia/Samland/Sambia or Žemaitija (Samogitish: Žemaitėjė; Lithuanian: Žemaitija), who along with Lithuania were one of the two core administrative units of the Grand Principality of Lithuania. Žemaitija is located in northwestern Lithuania.